T0332260

Stealing Time: Exploration in 24/7 Software Engineering Development

RIVER PUBLISHERS SERIES IN INFORMATION SCIENCE AND TECHNOLOGY

Volume 7

Consulting Series Editors

Prof. KC Chen

National Taiwan University, Taipei

Taiwan

Information science and technology ushers 21st century into an Internet and multimedia era. Multimedia means the theory and application of filtering, coding, estimating, analyzing, detecting and recognizing, synthesizing, classifying, recording, and reproducing signals by digital and/or analog devices or techniques, while the scope of "signal" includes audio, video, speech, image, musical, multimedia, data/content, geophysical, sonar/radar, bio/medical, sensation, etc. Networking suggests transportation of such multimedia contents among nodes in communication and/or computer networks, to facilitate the ultimate Internet. Theory, technologies, protocols and standards, applications/services, practice and implementation of wired/wireless networking are all within the scope of this series. Based on network and communication science, we further extend the scope for 21st century life through the knowledge in robotics, machine learning, cognitive science, pattern recognition, quantum/biological/molecular computation and information processing, biology, ecology, social science and economics, user behaviors and interface, and applications to health and society advance.

- Communication/Computer Networking Technologies and Applications
- Queuing Theory, Optimization, Operation Research, Stochastic Processes, Information Theory, Statistics, and Applications
- Multimedia/Speech/Video Processing, Theory and Applications of Signal Processing
- Computation and Information Processing, Machine Intelligence, Cognitive Science, Decision, and Brain Science
- Network Science and Applications to Biology, Ecology, Social and Economic Science, and e-Commerce

For a list of other books in this series, see final page.

Stealing Time: Exploration in 24/7 Software Engineering Development

Editors

Zenon Chaczko

Ryszard Klempous

Jan Nikodem

Routledge
Taylor & Francis Group

LONDON AND NEW YORK

Published 2010 by River Publishers
River Publishers
Alsbjergvej 10, 9260 Gistrup, Denmark
www.riverpublishers.com

Distributed exclusively by Routledge
4 Park Square, Milton Park, Abingdon, Oxon OX14 4RN
605 Third Avenue, New York, NY 10017, USA

Stealing Time: Exploration in 24/7 Software Engineering Development / by Zenon Chaczko, Ryszard Klempous, Jan Nikodem.

Routledge is an imprint of the Taylor & Francis Group, an informa business

ISBN 978-87-92329-42-4 (print)

While every effort is made to provide dependable information, the publisher, authors, and editors cannot be held responsible for any errors or omissions.

Preface

Stealing Time: Explorations in 24/7 Software Engineering Development edited by Zenon Chaczko, Ryszard Klempous, and Jan Nikodem focuses on how to efficiently and effectively develop software in the era of globalization. The realm of "24/7 Knowledge Factory" conjures up concepts such as Team-based, World-wide and Cross-time. We often ponder what the benefits or drawbacks of the continual work flow might be, what impacts on productivity, culture and society the new paradigms will have. The collection at hand attempts to address these issues from a specific perspective, namely that of software engineering.

This edited monograph contains seven chapters:

In Chapter 1, the authors propose Teaching Practice-based Subjects in 3 Time Zones Virtual Student Exchange Environment methodologies. They explore and evaluate a new collaborative framework for an e-learning system in higher education.

In the first part of Chapter 2, the authors explore the impact and implication of the Eventflow mechanisms in the 24/7 Virtual Student Exchange (VSX) environment. Then, they describe remote Virtual Machine Labs (VM Labs) for undergraduate students in large, team-oriented coursework setting.

Implementation of the "follow-the-sun" work paradigm as the 24-Hour Knowledge Factory, with which we can transfer knowledge across borders and cultures, is discussed in Chapter 3.

In Chapter 4, the author presents a case study on using Mass Spectrometry for cancer detection research (SELDI-TOF-MS) that takes an advantage of the development of testing preprocessing, classifying algorithms and software modules in the 24/7 continuous engineering mode by geographically distributed and cooperating teams.

An adaptive production process and organizational structure in World-wide Teams in Software Development is proposed in Chapter 5. This process solves the most important problems in the 24/7 software development, more specifically, it allows for a quick reaction to changes in execution, it mitigates the risk of incorrect functional requirement specifications, and facilitates the verification of the final product's quality.

In addition to core disciplinary technical skills, modern engineers must possess cross-cultural communication skills, team management skills, and the ability to perform effectively within geographically distributed teams. Chapter 6 explores one effort to integrate such training into the traditional engineering education paradigm by introducing a novel, internationalized curricular model that forces students to engage in the challenges of 24/7 engineering.

In Chapter 7, the authors propose Data and Knowledge-Transfer Model for the development of collaborative requirements analysis CASE tools designed for cross-time-zone development teams.

The book is a comprehensive collection of the most recent work in the field. Many readers will find it fascinating to discover the state-of-the-art discussions of the emerging, globally integrated software development.

飯島淳一
Junichi Iijima, Tokyo

Jerzy W. Rozenblit, Tucson, Arizona

Editorial

Rapid changes in information, communication and computer technologies have a dramatic impact on teaching and learning methodologies. Soon students in tertiary institutions will be able to work in geographically distributed environments and develop multi-lingual, multi-cultural communication skills and cross-disciplinary interests without ever taking part in a physical exchange. Because of globalization, distances between people are getting smaller and should be taken advantage of. These days, we deal in virtual spaces all the time and we have technologies that can support remote group-work methodologies, however, we need to develop people to complement these.

This book provides an overview of investigations into groupware technologies that aim at overcoming time, geographic and cultural differences by managing and engineering cross-institutional cooperation. This allow students at one educational institution to participate virtually in team-oriented and practice-based courses offered at other universities around the world. Exposing students to different educational institutions, different programs and ways of teaching and learning can be extremely useful for people with financial or physical inability to travel overseas.

Various projects and initiatives presented in this book are about sharing experiences, resources, teaching methodologies and tools. Learning in remote group-work environment enables students to be exposed to different social behaviors within different cultures and in a comfort of familiar environment. Authors of this book recognise the importance of new approaches to teaching and learning by investigating what tools and methodologies could be used to support the new global model of educational cooperation. Where suitable tools don't already exist they are helping to develop new frameworks to fill the technological and organizational gaps. While telerobotics, teleconferencing and remote monitoring and viewing already exist, now it's time to take this kind of human activity to a new level. With the increasing reliance on remote work and distributed teaming in the modern industrial contexts, educators around the world must work to adapt traditional engineering education models to provide the experiences and training required to succeed in this global work environment. The authors of this volume clearly feel that the answer to a globalized work environment is to globalize education as well, with discrete centralized knowledge repositories replaced by collaborating international education consortia. This approach maintains a certain autonomy among institutions, while promoting sharing of educational programs and experiences with minimal costs.

Editors of this book would like to thank Prof. Szafran, the VC of the Wroclaw' University of Technology, Poland for financial assistance in preparing this book. Also, we would like to thank Mr. Chris Chiu for helping us in refining the format and preliminary editing of the book.

Zenon Chaczko
Sydney, Australia

Ryszard Klempous and Jan Nikodem
Wrocław, Poland

Contents

**3   Toward the 24-Hour Knowledge Factory in
       Software Development   35**

Amar Gupta, Satwick Seshasai, Igor Crk, David Branson Smith

5   Worldwide Teams in Software Development   97

Pawel Cichon, Zbigniew Huzar, Zygmunt Mazur,
and Adam Mrozowski

6   Virtual Student Exchange: Developing New Educational Paradigms to Support 24/7 Engineering   125

Eckehard Doerry and Wolf-Dieter Otte

1

Teaching Practice-based Subjects in 3 Time Zones (3TZ) Virtual Student Exchange (VSX) Environment

Zenon Chaczko[1], Ryszard Klempous[2,*] and Jan Nikodem[2,†]

[1]*Faculty of Engineering and IT, University of Technology, Sydney, Australia*
 zenon@eng.uts.edu.au
[2] *Wroclaw University of Technology, Wroclaw, Poland,*
ryszard.klempous@pwr.wroc.pl, †jan.nikodem@pwr.wroc.pl

Abstract

Effective and successful e-learning occurs within a complex system that encompasses "the student experience of learning, teachers' strategies, teachers' planning and thinking, and the teaching/learning context" [1]. Education and research communities (at least in theory) are perhaps the most open and the least affected by negative aspects of human psychology often exhibited in collaborative environments. The proposed Teaching Practice-based Subjects in 3 Time Zones (3TZ) Virtual Student Exchange (VSX) Environment [2, 4, 6] methodology intends to explore and evaluate a new collaborative framework for the design, development and implementation of e-learning system in higher education. The project is compatible a model of teaching and learning involving a blend of three interrelated features: a practice situated in a global workplace, an integrated exposure to professional practice as well as a research inspired and integrated learning.

Keywords: Three Time-Zone (3TZ), Virtual Student Exchange (VSX), Collaborative Environment, Group Project Management, On-Demand Virtual Labs (ODVL)

1

1.1 Background

The teaching practice-based engineering in 3 time zones (3TZ) intends to address three main challenges for collaboration oriented educators creating and supporting educational environments in which students can learn to practice software design in teams. The first challenge that an effective communication is being managed by enabling students to experience and recognize difficulties when communicating and working in diverse set of teams, members of which might come from many diverse ethnic backgrounds, locations and teaching environments. The second challenge is that resource management is being practiced and skills being developed in extreme conditions where team members and project resources may not be collocated. And finally, the third challenge that orderly executed delivery of a high quality projects (assignments) in time constrained environment. In order to meet the above challenges we aim to perform the following activities and tasks:

- Develop our understanding of the situational contexts in which the innovative 3TZ model would allow to foster desired communication and management skills
- Design effective teaching and learning opportunities by adaptation of various project scenarios embedding cooperative behavior, facilitating new teaching and learning assessment possibilities in various engineering subjects taught in cooperating institutions
- Provide a study of how we could benefit from sharing teaching experiences and resources among participating institutions, such as use of remote labs and On-Demand Virtual Labs (ODVL) facilities [14].

Specifically it would be important to investigate how students perceive and reflect on teaching projects that involve cooperation on a global scale. Initially, the main focus should be placed on monitoring the quality of adapted processes, tools and interactions between project teams and individual students. The design of methodologies, tools and future projects is to be grounded by the integration of:

- Event-flow based, auto-reporting methodologies and techniques for collaborative project management (possibly to be adapted in many other disciplines) that arise from our recent research see [4, 5, 6, 8].

- A range of semi-automated and automated (web-based) project narratives (i.e. project logs [8], blogs [7], minutes and reports source from and related to the original auto-reporting, but recounted from various perspectives (i.e., global team members providing narratives with varying views that reflect their contexts).

The pilot project will bring together academics from three cooperating universities: Faculty of Engineering & IT at the University of Technology, Sydney, the Faculty of Electrical and Computer Engineering at the University of Arizona, Tucson, USA and the Wroclaw's University of Technology in Poland [4]. Strong research links have been established between cooperating academics in those institutions. Each member of the proposed project team is to embed the above aims and objectives in within her/his subjects thus enabling the benefits of various methodological adaptations across subjects. Participating institutions are located in 3 separate time zones each 8-hrs apart. Following the evolutionary Software Development Life Cycle model, the assessment process and the artifacts would then be redesigned and improved on, based on results provided through evaluation.

1.1.1 Benefits

It is expected the all participants would benefit from the auto-narrative models of assessment techniques for VSX developed during this initial project. Enriched education is a very important factor to our university and the society as a whole. The assessment techniques for learning and teaching in the diverse virtual environments are yet not an entirely explored theme. Significant benefits could be achieved by running collaborative projects across the globe due to different teaching schemes allowing a sharing of resources and cost cutting. This research project integrates well with the main initiatives such as laparoscopy simulation labs, sensor-net labs, remote labs and ODVL. The 3TZ VSX Environment approach can enhance these flagship research areas bringing such specific benefits cooperating universities including:

- Better utilisation of lab equipment, computers and software,

- Providing students and staff members international experience with no travel involved,
- Exposure to global and collaborative engineering practices and improvement in communication skills that could prepare students for the global market place,
- An enriched learning and teaching experience of students and academic staff.

1.2   Addressed Priorities by Faculty, Unit and University

The project addresses academic priorities associated with "continually improving quality" strategies undertaken in cooperating institutions. These involve:

- Building and improving the quality of teaching and learning: the project addresses priorities "better mechanisms of monitoring", "analysis and re-engineering processes" and "build on teaching and learning environment; working on innovative teaching methods and assessment".
- Designing new e-learning strategies and processes; the project addresses the priorities "analysis and re-engineering processes" and "measure our success".

The proposed method additionally addresses such objectives [13] as:

- "Increase our capacity to identify, recognize and reward effective teaching and learning" and "Improve supporting mechanisms for students and staff that enable effective teaching and learning."
- "Promote inclusive learning practices and infrastructure that recognize the diversity of students and their range of needs", "Ensure informed and effective implementation of technology, information and physical resources and infrastructure to improve learning, social and intellectual outcomes by students."

1.3   Rationale and Approach

Why 24-hour continuous development in 3 (separated by 8-hours) time zones? The idea of teams working in geographically and temporally spaced environment is not new. At present, many large companies like Motorola, Siemens, Canon, Nokia and Google have adopted the idea of working in collaborative teams across the globe. The Open Source communities (e.g. the Grid, Linux kernel, Apache server and others) have worked in this mode for decades. Their approach however, relies on efforts of a selected number of individuals or a leading team performing activities in one location and sharing testing or integrating tasks at various sites at the end of product development cycle. However, it comes as a shock to the majority of university graduates to work in a world-wide and global 24/7 business environment.

Therefore, the principles of cooperative study and work in teams are well understood and frequently practiced in subjects taught within engineering programs. Several research papers have been published documenting the educational model involving the 3TZ framework as well as discussing the e-learning and collaborative processes involved [4, 5, 6, 7]. However, studying, working and learning in a collaborative VSX environment in 3 separated by 8-hours zones is a new concept. The Wall Street Journal in 2007 reported on the project [11], indicating that universities may explore such a model of teaching. In a global market, there is a strong need for subjects to be taught and practiced following the VSX model within the curriculum. While there are many techniques and approaches for assessment of collaboration quality, currently there is limited number of studies reporting on the quality of experience and learning in distributed teams' setup as proposed in this educational project.

1.3.1 Project Methodology

The project applies an action research method. In the initial phase all students undertaking the project based undergraduate subjects (or postgraduate subjects) as a part of their individual assessment are asked to provide a logbook or an electronic blog in which they record their project activities over the entire semester. A specific format is required for keeping their records (e.g., dates, times, location, activity type, duration, aims, purpose, results, technical notes, diagrams, design and personal reflections). The logbook is then used by lecturers/tutors as one of the determinants of assessment of the quality and the level of student's participation. In the ICT and engineering industry keeping a logbook is a standard practice which often helps to enact project events or their flow. In a dynamic 3TZ VSX e-learning environment where students are coming from different countries and cultural zones — keeping a consistent format of student records and then effective access to these records for their evaluation pose significant challenges. In recent times there has been an increased interest in techniques that could capture not only individual records but the entire team's dynamics and its activities in a spatio-temporal and event-stream context (refer to work discussed in [8, 9]).

1.3.2   Learning

Development of e-learning in 3TZ is not just about having the focus on "the level of technological delivery strategies" but also about addressing important aspects of the teachers' conception of learning that influence the quality of "the planning of courses, development of teaching strategies and what students learn" [1]. Effective teaching requires an exchange of teaching ideas and collaboration that takes place while using the web and remote labs technologies beyond the bounds of a single university, the objective is to exploit the benefits and usefulness of this innovative approach for tele-collaborative teaching and learning in order to increase their efficiency and effectiveness. The VSX model [4] is being defined to explore efficient continuous modes of teaching and learning. This new methodology addresses the needs of research and industry communities to train in international teaming, facilitate research, enrich students' experience, and to improve the quality of collaboration (research/education) of the participating institutions. Prototypes of various models and technologies would be explored in order to create an integrated framework for international collaboration among teaming groups of researchers/students in order to practice team oriented education and research in a 3TZ VSX mode.

1.3.3   Embedding and Sustainability

How the project outcomes will be embedded in subjects/programs and sustained after the funding period? The project outcomes, a developed methodology, toolkits and a system prototype, will be used to analyze learning experiences and dynamics of students participating in VSX based projects and courses. The results of this analysis will be presented to subject coordinators. The obtained information will be used by subject coordinators, program heads, teaching & learning associate deans and the university student administrators for promotion/selection of new methodologies in the renewal of the teaching curriculum. In effect, the students will enrich their learning experience and gain unique communication skills.

1.3.3.1 Intended Evaluation of the Project Outcomes

The success of the project, with its short and limited timeframe, cannot deliver a full evaluation of the expected business effects, for that reason it will focus on validation of the 3TZ model and development of tools for evaluation of teaching and learning results. The project and the methodology will undergo formative evaluations of progress and outcomes. For each milestone, the project team will review the progress and results. There will be a final assessment and evaluation by subject coordinators and the teaching staff of the first trial of the new teaching model and related tools. The summative evaluation and assessment will take place by presenting a report on the results of the adaptation of the model to teaching staff and students. The project final outcomes will be presented at the teaching and learning forums. For the validation of technical performance, the project outcomes will be evaluated by: (a) selected evaluation metrics and key performance indices (KPI) will be developed to measure the significance of the identified teaching patterns and their impact on teaching and learning performance analysis; (b) coordinators/teachers of the selected subjects will be invited to participate in assessment of the outcomes and provide a feedback that will be used in further refinement of the 3TZ model; (c) a survey (questionnaire) based on the analytical results will be prepared and distributed to students to gauge their views and collect their feedback for further processing. The aim is to improve the identified patterns (reify anti-patterns) and verify the group dynamics and finally; (d) a random group of students will be tested and traced to verify the performance of the teaching model. In the above way, we will ensure the 3TZ method and the project is deliverable, effective, workable and sustainable.

1.3.3.2 What dissemination activities could be involved?

The dissemination may include the following aspects:

- Studies on the selected programming and database subjects taught at various engineering and IT faculties;

- Based on the above analysis results, a student profile and learning performance reports can be generated for the selected subjects and programs at various schools. The report might be circulated to the executive committee members, course committee members, quality committees' members and subject coordinators and T&L deans for evaluation.
- Final project reports should be produced describing project activities and outcomes to date and including a statement of expenditure against the project budget;
- the a piloting project or initial funding period, a workshop introducing the methodology and developed tools to the subject coordinators at various schools at the faculty forums. This should be organized to encourage lecturers to use the method in their subjects.
- Preparations should be made for broader evaluation of the longer-term outcomes and the impact of the project to all faculties in the cooperating institutions.

References

[1] S. Alexander, "E-learning Developments and Experiences", Education + Training Journal, vol. 43(4/5), 240–248, 2001.

[2] A. Gupta and S. Seshasai, "Toward the 24-Hour Knowledge Factory", Massachusetts Institute of Technology, 2004.

[3] E. Doerry et al., "Virtual student exchange: lessons learned in virtual international teaming in interdisciplinary design education", Proceedings of the Fifth International Conference on IT Based Higher Education and Training, ITHET'04.

[4] Z. Chaczko, R. Klempous, J. Nikodem, and J. Rozenblit, "7/24 Software Development in Virtual Student Exchange Groups: Redefining the Work and Study Week", ITHET 2006, Sydney, Australia, July 2006.

[5] Z. Chaczko, R. Klempous, J. Nikodem, and J. Rozenblit, "Assessment of the Quality of Teaching and Learning Based on Data Driven Evaluation Methods", ITHET 2006, Sydney, Australia, July 2006.

[6] Z. Chaczko and S. Sinha, "Strategies of Teaching Software Analysis and Design – Interactive Digital Television Games", ITHET 2006, Sydney.

[7] Z. Chaczko et al., "Blogging as an Effective Tool in Teaching and Learning Software Systems Development", Blogtalk Downunder, Sydney, May 2005.

[8] Z. Chaczko, J. D. Davis, and C. Scott, "New Perspectives on Teaching and Learning Software Systems Development in Large Groups-Telecollaboration", IADIS International Conference WWW/Internet 2004 Madrid, Spain 6–9 Oct., 2004.

[9] E. Freeman and D. Gelernter, "Lifestreams: A Storage Model for Personal Data", SIGMOD Record, 25(1), 80–86, 1996.

[10] P. Kandola, "The Psychology of Effective Business Communications in Geographically Dispersed Teams", Cisco Systems, 2006.

[11] A. Gupta, "Expanding the 24-Hour Workplace", The Wall Street Journal, September 15, 2007.

[12] UTS:IT Strategic Plan 2005–2008: http://wiki.it.uts.edu.au/admin/images/2/24/Faculty_of_Information_Technology_Strategic_Plan_2005-2.pdf.

[13] Setting the Pace 2006 -2009 Strategic Direction for the Current Decade: http://www.planning.uts.edu.au/pdfs/settingthepace.pdf.

[14] J. Lucas and S. Murray, "Teaching Operating System Concepts Through On-Demand Virtual Labs," Proceedings. of the 2008 AaeE Conference, Yeppoon, Australia.

2

Collaborative Team Project Management

Zenon Chaczko*, Christopher Chiu†, Michael James Lucas‡

Faculty of Engineering & IT, University of Technology, Sydney, Australia
*zenon.chaczko@uts.edu.au, †christopher.chiu@uts.edu.au,
‡michael.lucas@uts.edu.au

Abstract

The first part discusses managing software activities by using event-flows in a collaborative environment. Over the years the information and communication technology has become alienated from the physical environment. Centralistic models of computer systems architecture followed traditional social models of central software project management and decision making. A centralistic approach to management and application of software resources is still the predominant model in educational environments. In most computer systems today, large amounts of data are transferred from users or environment to a central processing place where information is created, stored and where decisions are made. This model carries not only much inefficiencies and contributes to serious problems in ever interdependent, interoperating systems and applications. By exploring the impact and implications of use of Eventflow mechanisms in 24/7 Virtual Student Exchange (VSX) environment, we realize how Eventflow facilities can assist communication and collaboration between culturally and time-zone diverse software development groups.

The second part investigates the application and effectiveness of deploying remote Virtual Machine Labs (VM Labs) for undergraduate students in large team-oriented coursework settings. While the use of virtual machines is not new for commercial enterprise developments, there is an awareness of the pedagogical benefits of using VM technologies for student activities in software project development. The student and teacher survey conducted at the end of semester indicate the benefits of VM technologies are not only reflected in terms of technological benefits in terms of system maintenance and a uniform platform environment, but to also enhance teaching practices, such as enabling students to dedicate more time to be spent in development work; and improved student-tutor feedback techniques as teachers can actively monitor the progress of development throughout the software development lifecycle.

Keywords: Eventflows, Collaborative Environment, Software Development Lifecycle, Remote Virtual Machine Labs, Virtualization, Pedagogical Methods, Group Project Development

2.1   Managing Software Activities in 24/7 Vsx Mode: Eventflows in a Collaborative Environment

2.1.1   Introduction

In the global economy, we have seen a decrease in the barriers towards communication across the globe along with an increase of service availability to support this communication. Software development is one discipline that is capable of effectively utilizing and benefiting from the global collaboration prospects lent by ever increasing capability of information and communication technology [17]. With the Internet software development teams do not need to be co-located within the same workplace nor even in the same time zone. Single members, small development and even large teams may be spaced across the globe, working on the same project in their usual working hours and passing information between different teams as they start or finish a day's work [4]. The Open Source community has, to a degree, worked in this fashion for many years. However, their approach usually involves either a core team(s) working in one location, or various individual developers working on separate tasks and integrating at the end of development. There is currently very little in the way of set engineering methods to facilitate this kind of development.

We shall investigate the methodologies, software tools for a telecollaborative software project management and a development of actor oriented development framework that can enrich the collaborative experience [3]. Initially, the investigation we will generate a framework of processes, which can be effectively applied to manage task and knowledge management between these time and geographically spaced teams. Where these will differentiate from current collaborative processes is in their ability to allow singular tasks to be worked on by separate teams. This framework will be evaluated through the cooperation of the various partner universities [6]. The feedback will be used to validate and verify the usefulness of the processes developed.

2.1.2   Problem Space

Large software development companies like IBM, Sun Microsystems, Cisco Systems, Nokia and Google have embraced the idea of working with geographically and temporally spaced teams. The Open Source community (For example the Linux kernel and Apache web server projects) has worked in this fashion for years, but their approach usually involve either a singular core team working in one location, or various singular developers working on separate tasks and integrating at the end of development. To effectively prepare university students to work in global 24/7 corporate environments [11], the principles of continuous collaborative development needs to be taught and practiced within the curriculum [14]. While there are numerous collaboration software platforms available, currently there is very little in the way of tools to facilitate the task management aspects of a distributed team educational project.

2.1.3   Why 24-hour Continuous Development?

24-hour continuous development is ideal for application towards tasks that have hard-deadlines or require work completed as soon as possible. If a functional or security bug is discovered in a mission critical application, there is a need to find a solution within the shortest period of time. A solution, which would normally take three days to work on, can be completed in a 24-hour continuous time period. A single site's employees cannot be expected to and would not have the capacity to, work continuously to meet the intensive activities involved, but a three site, 24-hour continuous development platform would meet these activities. For companies offering 24/7 systems support, this two day difference can mean the difference between an inexpensive and an expensive support cost.

Fig. 2.1 Working time available on single site.

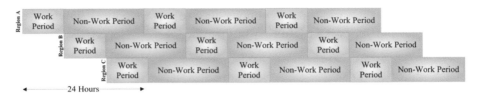

Fig. 2.2 Working time available on multiple sites.

2.1.4 Models of 24-hour Continuous Development

Different models exist for how best to approach the 24-hour continuous development problem. Some of these are the modular approach seen in the virtual student exchange study and the continuous project management structure seen in the study run by CaST at University of New South Wales [9]. Chaczko et al. [3] and Kandola [13] found that virtual groups work better and prefer to communicate in a task-oriented fashion.

2.1.5 Software Development Methodologies

Iterative and Incremental processes such as agile software development frameworks are ideal for continuous collaborative development. The small iterations of the project development already allow for periodic re-synchronization, these iterations can be utilized as a period where the project management team can ensure that the project vision is shared across sites. Any inter-site related issues arising in the previous iteration can be isolated and rectified before moving onto the next iteration.

Scrum is an agile software development methodology that overlaps nicely with the concept of continuous software development, an iterative process where the development phases are overlapped and performed by a cross-functional team. For example, the handover-synchronization components of continuous collaborative development can be achieved, from a project management perspective, in the daily team meetings that occur during a scrum. As

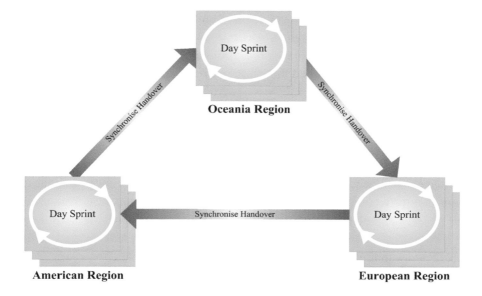

Fig. 2.3 Three-site SCRUM sprint.

the daily meetings only occur during a scheduled scrum, this synchronization would run smoother when dealing with intensive development tasks linked to scrums.

2.1.6 Eventflows

Software engineers are using an ever increasing number of Computer-Aided Software Engineering (CASE) tools to document or assist in all areas of software development. It would be useful from an organizational knowledge point of view if data produced (such as flat files) and modifications of this data could be captured and correlated against work done in other tools; however modifications that are made in these various tools are currently only recorded in the persistent data of the application. Decision making in distributed teams can also be fragmented and individual teams may need to make decisions that affect the entire project but which cannot be made in with collaboration or agreement of all teams. Where and when these decisions are made, and how they affect the project needs to be recorded and distributed to all distributed teams.

Here we introduce the concept of Eventflows, which are an evolution of the Lifestreams concept first written about by Freeman in 1996. Where Lifestreams record the digital events of a single person, Eventflows record the digital events of a project and the project artifacts. Eventflows capture events and periods within the project's global system and in-turn be used to capture and distribute project knowledge. Events can be classified as any significant occurrence on the project that can be captured or recorded by a computer, for example the login or logout of a system, the commit of changes to a version control system or the modification of a project artifact. A period is the linking of two key events where on their own has little or no value. Eventflows can be captured through automated systems or through manual creation from users. Eventflows can also be linked against individual or groups of tasks defined in project management tools such as Microsoft Project, thereby showing the actual work that was required and accomplished to complete the task.

2.1.7   Eventflow Specification

An eventflow must consist of both human readable and application readable data. In the most primitive version of an eventflow it must contain the following:

- Date and time of creation — Time is expressed in UTC.
- Human readable description of event — Expressed as a string object.
- Machine readable description of event — Implemented as serialized object.
- Project identifier — Optional for single project repositories.
- Artifact identifier — Unique identifier of this artifact in the project.

2.1.8   Eventflow Server

The Eventflow design is modular and can be implemented as a standalone library or as a separate server. A service-orientated architecture of the eventflow server will allow Eventflows to be integrated into a wider range of external collaboration and task management environments and provide further avenues to research into organization knowledge systems in software engineering [15]. Architecturally this would allow for greater performance for collection and

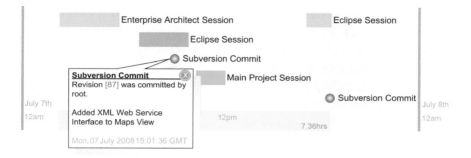

Fig. 2.4 A prototype of a single user eventflows for a day of a project.

processing of eventflow data. An Eventflow server would also aid in integrating external CASE tools such as Sparx Enterprise Architect.

As previously mentioned, an eventflow server can record all of the digital events in a project from distributed sources. Through data-mining and business logic analysis, decision making can be tracked through the project. Theoretically the server could make automated decisions or reports based on a project's eventflows. Figure 2.4 depicts a prototype of a single person's eventflows for a day of a project. In this example the eventflow shows that a user was working on changes within an Enterprise Architect project, then making changes to code in Eclipse and eventually committing the changes to subversion (with these last 2 steps repeated). The application readable components of the eventflow can identify where a modification to a component in Enterprise Architect was eventually committed as a modification to the code base of the application.

2.1.9   Closing Statements and Future Work

What we are ultimately trying to achieve is to exploit benefits and usefulness of this innovative approach for telecollaborative research, teaching and learning in order to increase the efficiency and effectiveness of these activities. The 24-7 VSX model is being defined, developed and applied to explore reliable and efficient continuous modes of research/study/teaching processes. The new methodology addresses needs of academic community and industry for training in international teaming, facilitates research, provides enriched students' experience and it improves the quality of collaboration in both research and education among the participating institutions. A new type of virtual lab will be equipped with advanced tools (hardware and software) in order to create an integrated framework for international collaboration among teaming groups of researchers/students in order to practice team oriented research/engineering and education in 24-7 VSX mode.

Specifically, the main thrust of our research is to build an experimental platform for theoretical development and experimental verification of 24-7 mode of telecollaborative engineering. The objective of the current stage is to build rudimentary collaborative network in which both human, computer and software actors could interact in a network environment that consist of remote labs, 24-7 VSX mode eventflow server, technology virtual machines and virtual portals (vortals). The ultimate goal is to setup a systematic design approach to solve various problems pertaining to applications described in aforementioned scenario.

We have finished the development of some low-level parts of the platform and conducted simulation and we have built eventflow server and vortal prototypes, but the objective is actually open-ended since we may have different high level missions/tasks defined. The key innovation is using the 'humans-virtual-hardware-in-the-loop' thinking in telecollaborative engineering research. Many researchers in collaborative teaching domain tend not to ask themselves many times of 'for-what' questions. We directly put the mission/task/activities in mind at the very inception of the project. We think the majority of education programs and the academic community will benefit from the ideas developed thanks to the 24-7 VSX model ultimately. Effective research these days requires multidisciplinary collaboration that often can occur in virtual-machines/remote labs environment. Enriched education is also very important for university communities and the society as a whole. Research into advanced technologies that can promote a better use and sharing available resource can be achieved among willing academic institutions around the globe.

2.2   Implementing Virtual Machine Lab Environments for Academic Group-Project Development

2.2.1   Introduction

The principal research goal is to investigate the experiences of deploying a virtual machine lab environment at the University of Technology, Sydney

(UTS), for the teaching and application in team-based software system development. The evaluation of student experiences is assessed with an end-of-semester electronic survey and their personal logbook entries, to understand the student's reflective process at each phase of the subject. Furthermore, the responses will serve to improve the application and deployment of VM Labs in future semesters.

This main observations derived from the surveys relate to previous experience in VM environments, the students' perception of usability of the technologies, perceived enablers of using VM labs and the demonstrable evidence of team productivity with the environment. At the final group presentations, it was expected that a core component of their prototype design would be implemented using VM labs, with discussions on student evaluation and feedback from the project stakeholders. In closing, a range of future research and development topics related to VM lab resources will be discussed as tangible evidence of students applying cognitive processes in their group work. It was anticipated that personal quality and the level of student participation enrolled in the subject would vary, however VM labs would form a key mechanism to offer effective sharing of information and opinions between peers in support of teaching, both from participant students and an educators' perspective. Combined with electronic web logs (blogs) [5], VM labs become an effective resource in tandem with traditional oral exams to assess and improve the process of teaching and learning in team-based subjects. Therefore, a key driver of VM labs was to improve cognitive, conational and emotive (CCE) [8] aspects of teaching and learning and be a contrivance for final subject evaluation and student feedback.

2.2.2   Group-Oriented Software Development Project Work

In the Information & Communication Technology Engineering course-stream (ICT), undergraduate students are assigned into teams who undertake to complete project work started within a requisite course, ICT Analysis (ICTA). The post-requisite course, ICT Design (ICTD), focuses on designing the system prototype to the customer:

- ICTA introduces methodical paradigms and requirements specification in the software development lifecylce (SDLC).

- In ICTD, a laissez-faire and constructivist approach [18] is achieved as knowledge is shared as a result of a team-oriented environment.

In the ICTA and ICTD project context, learning is a developmental process originating between people (socio-cognitive domain) and eventually within the individual (intra-personal domain). Teaching and learning in the subjects remain a constructive activity that sources new methodical approaches and emerging software technologies for team collaboration. The instructor's role in both ICTA and ICTD is foremost a teacher and project facilitator, but by ICTD becomes principally the final customer, moderating the individual effect of students exploiting a void within the dynamic of the team [20].

This activity is fostered by expecting students to form and manage their own groups, adopt their own methodological processes and paradigms taught in ICTA. Students become responsible for their own learning and a deliverable outcome, as the compulsory criterion is that participants must present a quality product that is on schedule and within budget. The underlying teaching and learning processes in ICTA and ICTD are fostering positive project development strategies that combine best practice methodologies and due process as defined by the project size and complexity [12].

The combination of the adapted activity focus method and spiral process model in the SDLC for teaching and learning in ICTA and ICTD is shown in Figure 2.5. The subject is facilitated by an ICT engineering team consisting

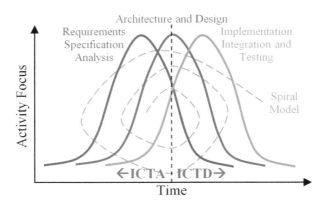

Fig. 2.5 Process activity method vs. spiral model for teaching and learning in ICTA/ICTD [2, 5].

of a coordinator, lecturer and tutor whose responsibilities include facilitating lectures, tutorials, and workshops, with conflict resolution support for complications that arise from group activities [5]. Final assessment consists of the group work project deliverables, engineering logs for individual assessment and an optional individual presentation of a self-assigned topic. Exit interviews are conducted for each team member to assess the relative contribution of each individual, and to assess the merits and approaches of designing their system prototype with VM Labs.

2.2.3 Application of VM Labs to Group Project Developments

Students perform their project development work by connecting to the VM Lab environment using a standard web interface. Seminars are provided to give an overview of how to setup and access their VM for their given project environment. Progressive functionality will enable lecturers to approve VM requests, and allocate resources as required by the students. The procedure for students to access VM Labs are as follows [1]:

- A nominated student requests for the operating system and development resources to be setup.
- An administrator approves and allocates a VM machine upon confirmation from the lecturer or tutor of the team.
- As shown in Figure 2.6, students can access their VM via the web interface to perform physical operations. Connection to the VM console is by Remote Desktop Connection (RDC).

The VM Labs infrastructure setup comprises of the core logical software components as depicted in Figure 2.7 [1, 10, 16]:

- A Virtual Infrastructure Layer where the VM Host Server is comprised of a bare-metal hypervisor. Virtual desktops are managed for student project development, containing a base operating system; and a resource allocation of memory, CPU cores and disk space. Port mapping can be activated for remote hosting of services such as databases and web servers.
- A User Connectivity Layer where users login via the web access server using their student authentication details stored in the LDAP server.

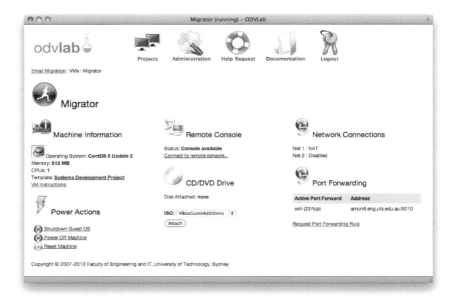

Fig. 2.6 VM Labs (odvlab) Web-Interface Front-end [1, 10, 16].

- Project deliverables and documentation are version controlled with the Subversion repository, with TRAC as the web project management system where users can post blog entries on project milestones and team progress.

The immediate advantage of this infrastructure setup is that academics have a global view of the development infrastructure that is setup for each team, and the management of the VM infrastructure is consolidated in a secure, unified environment.

2.2.4 Case Study 1: Visitor Monitoring System

The team assigned to the Visitor Monitoring System (VMS) was provided the project brief where a system would be designed to assist National Park and Wildlife Service (NPWS) rangers to monitor visitors perusing the national parks around Sydney. The system would be able to track visitor movements using a GPS Navigator, and respond to visitor distress or anticipate missing visitors in a managed park. The predictive element within the system is

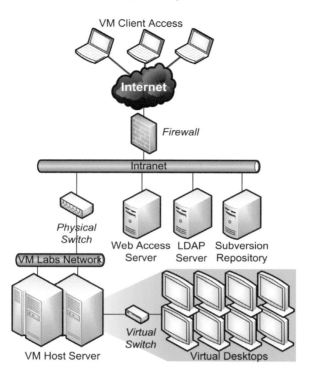

Fig. 2.7 VM labs infrastructure setup [10].

designed with the heuristics engine that would estimate missing user positions in a body of space.

The VMS teams used VM Labs to setup the Java development environment using open-source software, with Eclipse IDE utilized for the presentation and application tier. A separate VM was used for their persistence tier; with the open-source MySQL database management system (DBMS) used for building and normalizing the database structure. It has been noted that the VM Labs was used extensively for team collaboration and knowledge transfer, so domain experts could impart their experience with team members using VM Labs as a demonstration system.

The following system components were integrated using VM Labs using an 'n-tier' architecture [1]:

- **Presentation Tier:** Web Container: A virtual machine was used for the Java web container, Apache Tomcat. The Web container handles the web service facilitation between external actors of the system; such as sensors, visitors and administrators; and the VMS business logic tier components.

- **Business Logic Tier:** Web Components: On the same VM as the presentation tier, a Web Application development strategy was implemented using the Java Enterprise Edition Framework (Java EE). The business-level components written in Java were deployed on the VM for the operator interface, visitor interface and environment display. Struts interfaced the presentation tier (Java Server Pages) to the business logic (Java Runtime Applications).
- **Persistence Tier: Database Server:** On a separate virtual machine, the MySQL DBMS was used to manage the persistence layer. The decision was made to provide redundancy between the tiers, as the persistence tier is responsible for account records and visitor information statistics. The statistics were analyzed with dead reckoning heuristic techniques in the middleware services layer. To allow the VM hosting the application tier to communicate with the persistence tier, virtual port forwarding was enabled to allow the application logic components to have proper authentication privileges.

Students noted that the consistency of VM Labs has permitted a greater focus on design requirements, as less time is taken to install development environments on their personal computers. Additionally, the team has synchronized their VM test-bed development to the Subversion repository to simplify internal control processes. Unit and Integration testing tasks were partially performed online as development work could be performed at a remote location, such as from home or at the workplace during lunch or rest breaks.

Experienced team members were inclined to streamlined development using VM Labs, by simplifying project work deployment without being concerned about data backups and security constraints placed within the network infrastructure. As a result, student collaboration has been maintained online compared to previous semesters, as development work is achieved with VM Labs being a facilitator for coding and testing activities.

Project tasks and organizational planning in the development and testing phases of the assignment was also documented on a virtual machine by the

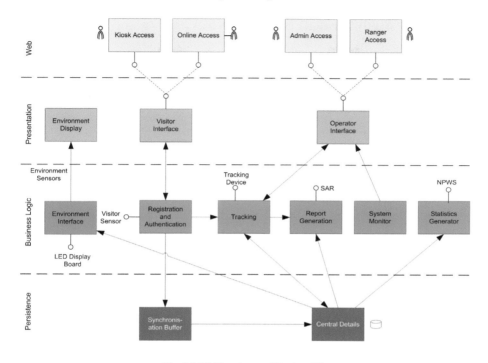

Fig. 2.8 VMS system architecture [1].

team leaders. By offering a seamless transition in the documentation control process, team meeting and project scheduling notes was done with VM Labs. While students kept complimentary written process logs, a number decided to write blogs with the TRAC project management system, offering transparency between team members' thought processes.

Team members decided to use the VM Lab environment as an active collaboration tool during team sessions. Furthermore, dependency on external hardware to install open-source applications and software resources was reduced, as team leaders would setup the environment in participation with the team to use. It is noted that while technical assistance was provided by the lecturer to setup and initiate the VM environment, there was no suggestion or ideas put forward by the teaching team as to how they should use the environment. In essence, VM Labs intuitively provided the means for team participation from the participation and engagement of students.

2.2.5   Case Study 2: Smart Hospital System

The Smart Hospital System team (SHS) was assigned to the task where a system would facilitate hospital doctors and nurses to manage patient medication with an automated robotic medicine dispenser. The teams were divided into two main sub-teams elaborated below:

- **Patient Information Management System (PIMS):** The PIMS team handled the hospital staff users and patient entry responsibilities. In addition, it manages medication dispensation duties that will integrate with the Tele-Medicine Cluster.
- **Tele-Medicine Cluster (TMC):** The TMC team formed the real-time robotic automation of medicine delivery and handling. It would be responsible for receiving the patient requests, checking and distributing the medicines.

For the PIMS team, the purpose of VM Labs was to use a virtual machine as a database server for their final prototype design. While it was intended for the team to design their database structure using VM Labs, due to internal team constraints the majority of design was completed on local computers. Final deployment of the persistence tier was merged on the virtual machine for use by team members. The PIMS system components implemented in VM Labs is a '3-tier' architecture [1]:

- **Application Logic Tier: Web Components:** A virtual machine was used to setup Internet Information Services (IIS) and .NET Framework, to deploy web services using Visual Studio 2008 Express Edition. IIS is integrated with Windows XP Professional, with IIS's web container facility hosting the web user interface. The presentation logic tier was implemented on the same VM as the application tier, including the Active Server Pages (ASPX) and C# code-behind
- **Persistence Tier: Database Server:** On the same VM as the application logic tier, the PIMs team decided to use a MySQL DBMS for staff user accounts, patient information and medicine prescriptions. While initially completed on personal computers, final deployment was achieved on the VM infrastructure.

The core development team members had previous experience to virtual machine environments, so technical inexperience was not a key factor in VM Labs being underutilized for the PIMS team. Apart from the technical constraints that were rectified over the semester as a result of continuous student feedback, students achieved limited collaboration work on the persistence tier using VM Labs. The main reason was a result of scheduling and organizational conflicts of the PIMS team, leading to development work being achieved late in the academic semester. Therefore, the initial preparation work to setup their VM environment for the team to use was not achieved, and thus the majority of development work was completed using their personal computers.

The TMC team comprised of designing the supervisory control elements of the SHS System. Team dynamics were hampered by a management structure that encountered social anti-patterns, including group-think and the false prophet syndrome. Technically capable team members with the domain expertise in VM Labs did not impart their skills to the rest of the team due to time constraints in developing the Citect SCADA data acquisition and control mechanisms. The only system components that had to be developed locally were the real-time elements in the system, including SCADA, vision detection and robotic and conveyor control. While there was initial interest to develop the persistence elements using VM Labs, ultimately it did not come to fruition.

In comparison with the first case study, upon first impression it can be inferred that VM Labs did not achieve its core objective of providing a resource fostering active collaboration. A further point must be noted that as with the first case study, VM Labs was offered as a repertoire of software support solutions that the students could use, but the manner in which it was to be used was never dictated — as to observe the effects of usage patterns for this pedagogical study.

However, a consideration of the team dynamics of both case studies reveals that in both instances, VM Labs is only a tool in a suite of academic resources available in group project environments. Teams that suffer from negative structural deficiencies such as individual effects [20] will still need active intervention on the part of the assessor, but verifying student participation by checking VM Labs logs ensures an additional degree of student accountability when it comes to exit interview assessments.

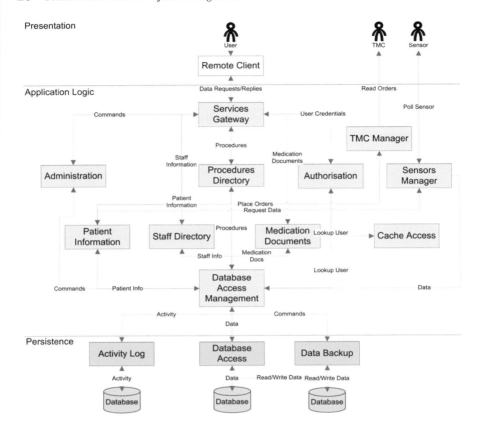

Fig. 2.9 PIMS system architecture [1].

2.2.6   Survey Analysis on VM Lab Deployments

2.2.6.1   Analysis of Student Survey Responses

The main observation from both case studies suggest students with existing familiarity with virtual machine environments, either from their industrial experience or previous academic or technical skills, have an discernible benefit of using VM Labs. Procedural administrative activities, such as requesting technical support to grant privileges to install the base operating system and integrated development environments, are no longer an obstruction to development. This is because these operations will be automatically initiated when a virtual machine request is made at the beginning of future ICTD coursework.

Another main benefit of VM Labs is the flow-on effect of domain-knowledge transfer, particularly from a technical perspective. Early adopters of VM Labs can drive early the development cycle to the team members who require greater software development and programming expertise. Students involved in the pilot study, intrinsically recognize that VM Labs provides a range of opportunities for collaborative teamwork, especially when it comes to the development, deployment and testing on a common clean test-lab environment. Apart from ensuring consistency when running uniform system tests according to specified standards, it also imparts best practice methodologies in the prototyping stage of development [18].

Future enhancements as a result of survey participation include offering multi-user VM access to enable team members to access the common VM environment to complete their assigned development responsibilities [1]. This reinforces group collaboration eorts as it can serve as a potential teaching environment for team members to tutor fellow team members by example. While personal responsibilities of students will lie solely with their individual preferences of telecollaboration, initiative on the part of team leaders will be imperative to foster this alternative approach when it comes to due-diligence and fulfilling personal obligations.

2.2.6.2 Academic Assessment of VM Lab Deployment

A common response from academics is the immediate benefit in terms of risk mitigation strategies when planning any team-based software project. As batch-script data backup responsibilities are automatically scheduled with VM Labs and the Subversion repository, the risk of data loss can be mitigated. The reasons can vary among the team dynamics and personal circumstances; it can be either genuine or a result of students exhibiting passive-aggressive behavior by maliciously corrupting data to provide time extensions to their work [8]. In combination with the Subversion version-control repository and TRAC web project management, tracking and control of group collaboration projects becomes easier for students and tutors. Apart from students learning how to effectively use revision control management tools in project development, the wiki personal logs allow students to share ideas and thought processes [5]. In essence, it offers a form of online brainstorming, where personal learning and development goals are shared on a common medium.

Subject coordinators and tutors can verify the personal ethics of students, as VM Labs access logs are archived to verify the student's individual contribution is achieved to the best of their ability. This is especially the case to instill that students need to be accountable for the efforts of their team, without resorting to unsubstantiated reasons to disown responsibility of assigned commitments when their team project work suffers from poor team dynamics. Therefore, there is an impetus for students to contribute in team environments for greater visibility and transparency in assigned responsibilities [2].

To summarize, the principal stakeholders of VM Labs get an opportunity to achieve the following goals to improve the pedagogical practice in team-oriented subjects [2, 5]:

- Define new subject requirements from a range of alternatives. Deficiencies in existing subject methodologies are addressed alternate proposals via blogs or student survey participation.
- Design proposed methodologies for the subject. Software and hardware resources can be laid out in advance for coming semesters to increase development times.
- Develop selective and efficient procedures for the subject, upon thoroughly testing these proposals. The training and experience of stakeholders with domain experts will enable future proficiency.
- Place the prototyping philosophy process with VM Labs into use. This is achieved by phasing in the new methodology to gradually replace the old methodology. The cost-effectiveness of a gradual or instantaneous transition depends on the particular situation.
- Evaluate the prototyping philosophy for the subject semester. Monitoring this methodology must be maintained consistently, with stakeholders updated concerning the latest modifications and procedures to how VM Labs should be deployed in future subjects.
- Recognize and evaluate the existing problems in team dynamics intelligently whilst performing some teaching and learning tasks. This is especially constructive in the assessment process by identify deficiencies early in the semester, and also during exit interviews.

2.2.7   Future Research Work

Over the course of this research study, there has been an apparent deficiency of useful general-purpose models available for problem-based [19] teaching when putting into practice collaborative software tools in the academic environment. The main issues related to learning and teaching of complex software systems development that still need to be addressed are [2]:

- How can teaching and learning models be integrated with VM Labs optimally? If not, how can it be evolutionary?
- How can constructivist models [20] be adopted for learning and teaching in large groups with the use of collaborative tools such as VM Labs, and how to implement these models in a subject?
- What changes are required in the academic program for instructors to deliver quality subjects that prepare students for a team project experience with the use of teamwork software?
- How can the school assist and benefit from this innovative pedagogical approach with VM Labs to teach team-based software systems development?
- How can the achievements of this pilot study lead to the development of new pedagogical resources suitable for ICT engineering programs that include systems analysis and design curricula?

2.2.8 Closing Statements

The research into implementing new teaching and learning methodologies is in its early stages, so naturally it is expected that the body of knowledge will expand over time. The experiences of using a collaborative development environment, such as VM Labs, exemplify ideas in academic practice that has the potential to improve interpersonal communication skills in student group projects. In essence, VM Labs is more than just an operating system virtualization environment; as the technological learning curve of VM Labs is more than compensated by the advantages in student productivity and reinforcing best practice through visibility and transparency in work outcomes. The impetus to involve students to engage with VM Lab environments is to create the awareness among tutors and students of utilizing a common hub to work in partnership together.

The proposed methods and project methodologies suggested only provide a glimpse of how team project development and ensuring knowledge sharing can be achieved in academic project work, beyond the realm of ICTD. VM Labs and the Subversion repository is not proposed as the complete technological solution to assist in fostering team dynamics, as can be found with the case studies. In reality, the current research initiative is designed to show the need to engage with subject stakeholders and students early in the project inception, and by driving collaborative technologies such as VM Labs in advance of system prototyping, such technologies can be viewed as an enabler to project development. The results of the pilot study should assist stakeholders involved in extending project management practices, with future technical work focusing on challenges to adapt to evolving technological advancements in software technologies for ICTD. The rapid prototyping model with VM Labs will form a part of specific scholarly approaches to create a new ICTD resource for student collaboration.

By providing a conduit within theoretical and practical approaches in student collaboration techniques, educating large groups of students in a team environment can be fostered through effective knowledge transfer and inspiring due diligence among team members. Even accounting for the complex characteristics that are evident in social team dynamics, the application of VM Labs with the relevant teaching and learning issues form an essential part of how the technology can be made seamless in future semesters [2, 5]. The techniques used to instill a collaborative development atmosphere in student team environments will provide insights into how VM Labs can be harnessed in student project development work. Continued research in this area will include implementing the suggestions of the pilot study in comparison with the original VM Labs platform, which will serve assessable results over a breadth of project scenarios in a range of ICT engineering disciplines.

Acknowledgements

We would like to acknowledge the contribution of the following people for their survey participation and participation into the research topic: the ICTD tutor Perez Moses and the students of ICTD.

References

[1] Z. Chaczko, C. Chiu, and J. Lucas, "On Demand Virtual Labs Support Survey", http://odvlab.eng.uts.edu.au/, Last viewed 30 November 2008.
[2] Z. Chaczko, D. Davis, and V. Mahadevan, "New Perspectives on Teaching and Learning Software Systems Development in Large Groups", IT Higher Education & Training 2004, Istanbul, Turkey.

[3] Z. Chaczko, D. Davis, and C. Scott, "New Perspectives on Teaching and Learning Software Systems Development in Large Groups — Telecollaboration", IADIS International Conference WWW/Internet 2004 Madrid, Spain 6–9 October 2004.

[4] Z. Chaczko, R. Klempous, J. Nikodem, and J. Rozenblit, "24/7 Software Development in Virtual Student Exchange Groups: Redefining the Work and Study Week", ITHET 7th Annual Conference Proceedings, pp. 698–705, 2006.

[5] Z. Chaczko, V. Mahadevan, and E. W. Chaczko, "Blogging in Teaching and Learning Software Systems", IT Higher Education & Training 2005, Juan Dolio, Dominican Republic, 2005.

[6] E. Doerry, R. Klempous, J. Nikodem, and W. Paetzold, "Virtual Student Exchange: Lessons Learned in Virtual International Teaming in Interdisciplinary Design Education", Proceedings of the Fifth International Conference on IT Based Higher Education and Training, ITHET 2004.

[7] E. Freeman and D. Gelernter, "Lifestreams: A Storage Model for Personal Data", SIGMOD Record, 25(1), 80–86, 1996.

[8] B. Ganley, "Blogging as a Dynamic, Transformative Medium in an American Liberal Arts Classroom," Middlebury College, USA, 2004.

[9] I. Gorton and S. Motwani, "Issues in co-operative Software Engineering using Globally Distributed Teams," Information and Software Technology, Elsevier Science, vol. 38, pp. 647–655, 1996.

[10] D. Grobler, "Virtual Desktop Access Proof-of-Concept," http://wiki.sun-rays.org/index.php/VDA_Cookbook, Last viewed 8 December 2008.

[11] A. Gupta and S. Seshasai, Toward the 24-Hour Knowledge Factory, Massachusetts Institute of Technology Press, 2004.

[12] P. B. Hastings, Blogging Across the Curriculum: A Course Resource for the Interactive Digital Design Program, Quinnipiac University, 2004.

[13] P. Kandola, The Psychology of Effective Business Communications in Geographically Dispersed Teams, CISCO Systems, 2006

[14] R. Klempous, H. Maciejewski, and J. Nikodem, M. G. Gonzalez, and P. Suarez, Informatics engineering curricula in the European Space Higher Education Area - towards a BAMA model, ICSE 2003. Coventry, 9–11 September 2003.

[15] J. Lucas, 24-Hour Continuous Software Development: A Framework for Use in Education, Undergraduate Thesis, University of Technology, Sydney, 2007.

[16] J. Lucas and S. Murray, "Teaching Operating System Concepts Through On-Demand Virtual Labs," Proceedings. of the 2008 AaeE Conference, Yeppoon, Australia, 2008.

[17] V. Mahadevan, Z. Chaczko, and R. Braun, The Telecollaboration Spin as a Concurrent Paradigm Shift in Business Practices, International Business Research Conference, Victoria University of Technology, Melbourne, Victoria, November 15–16 2004.

[18] I. Sommerville, Software Engineering: Practitioner's Approach, 6th Edition, Addison-Wesley, 2001.

[19] C. Warhurst, "Developing Students' Critical Thinking: the Use of Debates", M. Walker (Ed), Reconstructing Professionalism in University Teaching, Open University Press, 2001.

[20] U. Wilensky and M. Resnick, New thinking for new sciences: Constructionist approaches for exploring complexity, San Francisco Press, CA, 2005.

3

Toward the 24-Hour Knowledge Factory in Software Development

Amar Gupta[*], Satwik Seshasai[†], Igor Crk[‡], David Branson Smith[§]

The University of Arizona, Tucson, United States of America
[*]*gupta@eller.arizona.edu,* [†]*satwik@mit.edu,* [‡]*icrk@email.arizona.edu,*
[§]*smith.davidbranson@gmail.com*

Abstract

The 24-Hour Knowledge Factory posits the implementation of a follow-the-sun work paradigm based upon knowledge transfer across borders and cultures: the 24-Hour Knowledge Factory. Recent history has provided us with an applicable proxy that will help today's business leaders better visualize the far-reaching implications of this precedent: a notion emphasized during the latter half of the 18th and much of the 19th centuries, that "The Sun never sets on the British Empire," came to exemplify the British Empire in all of its far-flung glory. Specifically, this notion was meant to convey that the empire was so vast and spread-out that the sun was always visible by some residing in the territories controlled. Although the British Empire, as well as its perceived physical monopoly of the Sun's light, has gradually disintegrated, an equivalent notion can be coined to describe our vision: "The Sun never sets on the 24-Hour Knowledge Factory!"

Keywords: 24-Hour Knowledge Factory, Software Development Lifecycle, Distributed Work Environments, Decision Rationale Module, Decision History Module

3.1   Introduction

The roots of the 24-Hour Knowledge Factories can be found as far back as the Industrial Revolution. Before the advent of work standardization (in this case, the dissociation of 'parts' from 'finished goods'), when final products were made from beginning to end by one artisan, each item was considered to be a singular piece of art — that is, each product was unique. By breaking production tasks down to simple, componentized activities, productivity attained new heights as artisans became employees and specialization abounded. For example, a gun could be produced by a logger, a blacksmith, and an ammunitions producer, rather than have all three functions performed by one person.

The advent of electronic computers, coupled with help from diminishing telecommunications costs, allowed for the establishment of multiple "factories" in differing time zones, transcending physical barriers. This is exemplified by the notion of 24-hour call centers. Call centers active during normal daytime hours in the respective countries can have calls automatically routed to their locations. Utilizing the earth's natural rotation, we can ensure that employees of call centers can respond to calls that come during normal daytime work hours applicable to their respective locations by creating 3 to 4 call centers in time zones that are 6–8 hours apart from that of neighboring call centers. This concept of multiple support centers has been gradually adapted to support global communications networks. A growing array of applications makes feasible a geographically distributed workforce of highly trained professionals who work in succession to complete an endeavor much faster and with less worker fatigue and lifestyle change than the classical scenario in which all work was performed at a single location.

Advancing this idea, the authors have found examples of microchip engineers designing chips from disparate locations around the world. Contrary to initial reservations about potential knowledge transfer complications and an increased complexity of scheduling and project management, this work structure promotes an efficient design process with a faster turnaround time. High-talent designers who might otherwise be forced to work at odd hours of the night may now work during normal daytime hours in their own countries, avoiding the "graveyard shift" and the accompanying fatigue. The implications of such a work structure are great: many industries may be changed through the utilization of professional service teams that are not limited by the traditional geographic and temporal constraints. We believe that this paradigm will dramatically impact businesses, specifically in the way that they build, test, sell and support their products and services. Years ago, the difference in time between the United States and India, for example, was thought of as a hindrance that would severely impair the ability of Indian firms to work with US counterparts. Now, however, many companies are realizing the benefits that the time difference between locations affords to various projects.

The practice of "offshoring" is an inherent element of any 24-Hour Knowledge Factory endeavor. Currently, offshoring occurs primarily as a cost reduction scheme. Over time, the ability to attain dramatically faster time-to-market (TTM) will be one of the key drivers of offshoring endeavors. The running example in this chapter revolves around software development — specifically new product development — to highlight the applicability of this new paradigm to complex and dynamic efforts. Efficient information management is the key to incorporating 24-Hour Knowledge Factory concepts in such development efforts.

This chapter uses a case study to highlight a 24-Hour Knowledge Factory model with integrated data analysis. While this study involved two sites within IBM, the findings and methods can be readily applied to endeavors that use three or more geographically dispersed sites within any corporation or across multiple collaborating companies. As compared to traditional single-site operations, significant differences were observed in the information sharing, collaboration, and innovations in work operations activities of the distributed teams. The quantitative measures used in this case study gauged data on aspects such as frequency and methods of collaboration, social and technical networks, and differences in handling strategic and tactical decisions. The qualitative parameters were elicited through interviews in which the stakeholders described their perceptions of the quantitative data, and their motivations for decisions related to knowledge sharing. The primary emphasis of the field study was to evaluate the role that spatial and temporal differences play in the creation of new software products, with such analysis serving as the foundation for studying the 24-Hour Knowledge Factory paradigm.

3.2 24-Hour Knowledge Factory — The New Work Paradigm

The term "24-Hour Knowledge Factory" implies a global work model in which members of a team — who are located at multiple stations around the world — work on a project around the clock; each team member works during normal workday hours within his or her time zone, and transfers the task at the end of such a workday to a second team member located in a different time zone, who then continues the same task. This enables a type of a shift-style work relationship between workers, one analogous to that originally conceived of in the manufacturing sector during the Industrial Revolution, but with applicability to knowledge-based work rather than pure manufacturing. As stated earlier, a globally distributed 24-hour call center is the simplest manifestation of this paradigm. While call center work is performed on an ad-hoc basis, a better example of a 24-hour pseudo-factory paradigm involves groups working together to accomplish a given set of deliverables, such as a software project, and transcending conventional spatial and temporal boundaries in the process of doing so [1, 2, 4, 58].

Software development usually entails a work style denoted by the transmission of knowledge between members of a development team to create a product. Figure 3.1 shows a conceptual knowledge factory with distributed software design operations located in three different countries around the world. In a

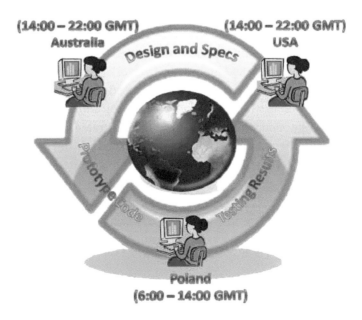

Fig. 3.1 24-hour tri-foci scenario.

delivery model such as this, a different, unique task is assigned to knowledge workers at each geographic location. This has the effect of improving the overall efficiency of the project, as workers in each location perceive that progress is made "overnight" while they are asleep. As will be discussed in following sections, additional models may be more applicable to different distributions of tasks, and tailor-made knowledge factories can address specific needs in diverse situations.

A successive, sequential, and circular work system is envisaged in such a "24-hour software development environment"; each distributed team will concentrate its efforts on a particular stage of the development process — designing the software, developing code, or testing of a subsystem — and each center will retain ownership of the endeavor for an 8-hour period in every 24-hour cycle. The ability to apply this paradigm to many other industries outside of the software industry exists, as many daily functions in multiple industries — accounting, consulting, and law, to name a few — employ a development cycle heavily reliant on sequential performance of specific functions. The environment that software development has traditionally relied on, an environment in which all parties are collocated, has led to an inefficient time lag between successive operations: engineers must wait for code to be developed before they can test it, and product developers must wait for a fully functional portion of the product before being able to take up production. By utilizing the earth's natural rotation, developers now have an unprecedented opportunity to create the product on an incremental basis, as they can now receive testing feedback overnight.

3.3   24-Hour Knowledge Factory — Foundations and Frameworks

Existing models that study strategic global plant location, primarily those employed in the manufacturing industry, have some applicability to the 24-Hour Knowledge Factory. Frameworks modeling the organization of decision making structures have been applied to plant location models, bringing them into the domain of knowledge based products and services. Loosely speaking, the analysis of the factors that play into strategic global plant location decisions has led to the determination of criteria appropriate for deciding plant locations for processes that involve the production of knowledge products such as software. Once geographical location decisions have been made, one can start to look at how to incorporate organizational decision-making and knowledge management frameworks in order to operate the globally disparate plants as hubs within a seamless and productive supply chain.

The 24-Hour Knowledge Factory has been researched in terms of its relevance to global production efforts before; the emphasis on knowledge as the key element being produced and transferred within this paradigm has been overlooked, however, and this is a seminal issue that is being highlighted in this chapter. As stated earlier, software development is a very good example of this issue; however, any other knowledge-based domain can be leveraged in a very similar manner. We have seen how previous research on optimal plant location for global manufacturing can be applied to the knowledge manufacturing domain; many other concepts can be seen as valuable building blocks of the knowledge-factory paradigm. Additionally, the growing phenomenon of offshoring of work and the organizational issues surrounding this function provide the basis of the "hybrid" model discussed in this chapter — i.e., a shared sourcing paradigm rather than one that involves complete offshoring of an entire business process.

3.3.1   Global Manufacturing — Plant Location

Lovelock shows that the services arena can benefit from lessons taught in manufacturing examples by understanding location choice and other global drivers [9]. Rosenfeld et al. have highlighted the need to consider parameters other than just cost in plant location decisions — local skill sets and other strategic considerations must be analyzed when attempting to create more flexible and efficient plants [10]. Manufacturing plant location choices can be delineated in a capability-focused model, with importance being placed on three decision-making factors: complexity, diffuseness, and well-developed interfaces [11]. Locations should be viewed as centers for value-addition that can be specialized to add value of a particular kind — this value must be identified in strategic plant location decisions [12]. Four sources of value exist: service, investment, cost, and profit. The 24-Hour Knowledge Factory specifies value-center specialization in each distributed location in order to add value to different areas of the supply chain in product creation. New product development requires that each of these value centers be present, furthering the applicability of the 24-Hour Knowledge Factory to dynamic challenges such as those encountered in the development process. Information management is another critical factor — one that spans the entire product lifecycle — in globally distributed development processes [13]. Overall, several factors beyond cost-cutting must be considered in the potential use of the 24-Hour Knowledge Factory paradigm as the vehicle to transform traditional manufacturing processes to focus on long-term success [14]. Finally, Pisano has displayed the need for investment in the manufacturing process by knowledge-based companies as opposed to traditional focus solely on R&D with manufacturing being outsourced [15]. Coming back to the running example of this chapter, software development is characterized by a 'manufacturing' process that is unique: the individual workers actually manufacture the product! In this sense, the 24-Hour Knowledge Factory can be seen as an innovation to the traditional manufacturing process.

3.3.2 Offshore Outsourcing

Agrawal et al. describe a phenomenon in which some firms are paying premiums to offshored vendors for "round-the-clock shifts", in which odd-hour work is performed to supplement processes performed in the waking hours of the geographic location of the customer [16]. Despite the premiums paid, cost reductions of between 30 to 44 percent can be achieved — even for complex, integral processes such as R&D — by sourcing work to these "round-the-clock" shifts. This model can be augmented by distributing work to a third location, and obtaining the benefits from the day-time shift style that bypasses the inevitable worker fatigue associated with night-time work. Kaka has presented a 6 model offshore partners schema that can be considered when choosing a model: supplemental staff, turnkey projects, assistance in building centers, build-operate-transfer, assets, and joint ventures [17]. The 24-Hour Knowledge Factory model can be viewed as the next natural step in this progression.

Research in the domain of offshore outsourcing has seen recent focus on emerging items that warrant a need for geographically distributed work. Saunders et al. show that, greater in importance than cost savings, technical capability is a very major consideration which, along with other core functions, should be maintained onshore to preserve technical capability [18].

Transaction cost economics have been cited by Barney as the only factor in determining the retention of tasks within company boundaries [19]. Carr suggests task modularization cannot be the sole focus of offshoring activities; on the aspect of building strategic competencies in all locations for all tasks must be considered [20]. Christiansen cites three factors that always exist for competitive advantage: economies of scale and scope, integration and non-integration, and process-based core competencies [21]. Semantic issues come into play as well, as highlighted in Light's emphasis on the broad understanding that managers must exhibit: both of their own cultural values and those of their employees in all locations [22]. DiRomualdo and Gurbaxani suggest guidelines on issues such as improving information systems, assessing business impacts, and generating new revenue when evaluating outsourcing [23]. Additionally, with respect to the widespread feelings in the U.S., especially the reluctance of its citizens to give up the high standard of living, Young states that individuals and companies in the U.S. must try harder to compete globally [24].

Progression of the notion of the 24-Hour Knowledge Factory will lead to a need for understanding the organizational issues faced by offshore workers, vendors and customers. According to Mizoras, the dynamics of the workforce and the notion of rework makes outsourcing challenging, leading to a need for third-party firms whose competencies lie in the realm of pure organization building practices [25]. One example of this is the notion of X-engineering as described by Champy; this allows for greater transparency, standardization, and harmonization, and results in dramatic alteration of the organizational relationship models [26]. Organizational learning is a critical factor in the building of flexible offshore models [27]; specifically, the focus on continuous, long-term learning as described by Lacity et al. [28]. They posit that a focus on efficiency rather than costs should occur when deciding where outsourcing should occur [29], and that factors such as the maturity of technology in these locations should be a critical factor in deciding what to outsource [30]. Quinn, in a fashion that echoes the notion of the 24-Hour Knowledge Factory, suggests outsourcing to strategic locations based on core competencies encountered at each location. In the model illustrated in this chapter, by allowing each task in the organization access to each location, relative competencies at particular locations can be exploited to their maximum potential, in a manner similar to the strategy expounded in theories of comparative advantage [31].

The alignment of product-market focus, organizational culture, resources and capabilities, and direction is noted by Fuchs [32]. The merging of these different foci is a great challenge, one that must be faced with many global centers, particularly when trying to mirror a one-organization model through working on a successive basis on an array of similar tasks. Cultural differences between knowledge workers are exacerbated when work on a particular project is shared. A GLOBE study confirmed that proper management of global teams requires an appreciation for specific cultural drivers that exist and are different across individual cultures [22]. Different treatment needs to be afforded to emerging markets in particular in terms of marketing strategy; this can be viewed as a cycle: markets will continue to develop, and learning will be borne of this development, leading to changes in strategy [33].

The 24-Hour Knowledge Factory revolves around a 'multi-shoring' concept, in which multiple locations are utilized as offshoring destinations based on different skills subsets and competencies in each location. Multi-shoring, however, does not consider the sharing of tasks between multiple locations. Millman has advocated enhancing outsourcing agreements so that processes are changed, a move well beyond the traditional 'commodity outsourcing' phenomenon seen in BPO offshoring today [34]; this serves as another confirmatory factor in terms of the necessity of outsourcing to the 24-hour model proposed here. Cheifetz advises strategic outsourcing for companies, based on a strong awareness of cultural differences [35].

Value chain and optimal process outsourcing decisions have come under scrutiny by researchers, and conflicting advice has emerged from this debate. Chesbrough states that inherent conflicts of interest between countries do not bode well for the outsourcing of innovation-based practices, and that negative ramifications would ensue [36]; whereas Quinn posits that competitive success is dependent upon outsourcing innovation — strategic outsourcing can allow for exploitation of talent at each stage of the value chain, allowing for new product development to be done in offshore locations [37].

Understanding within what "ecosystem" the organization exists is of seminal importance, as is the analysis of the health of those counterpart organizations that will have an effect on firm performance [38]. A 24-Hour Knowledge Factory that focuses on the software development environment provides a robust 'ecosystem', as this concept applies to the team as well as to the organization as a whole. Working in this manner involves a global mind-set, unconstrained by single country concerns or cultural factors while making business performance decisions [39].

Kumra states that customer-proximity activities need to be distinguished from remotely performed activities, and that a proper distribution of these tasks must be made [40]. The 24-Hour Knowledge Factory model allows for both customer-facing and remote components with the added ability to leverage both skills at any one location, as these tasks are shared. In a complex environment that can quickly change from initially expected processes, relationship management is critical to effective outsourcing ventures [41].

Finally, outsourcing requires management based on results negotiation rather than order issuance [42]. In a 24-Hour Knowledge Factory, the negotiation of tasks needs to occur at the team level — just as it would within a collocated team — since the varying degrees of required coordination disfavor centrally managed systems.

3.3.3 Information Technology Management

Taylor and Bain term call centers as "white collar factories" and have found employment relations to be analogous to ones in the manufacturing factories [43]. Such tight interaction between employees as necessitated by the 24-Hour Knowledge Factory means that disparities in employee relations between sites have the possibility to manifest as significant issues. Knowledge-based economies such as the United States are driven by talent; thus, management of talent should not be outsourced. Rather, talent management should be treated as a necessity for success [44]. Aron and Singh have shown that IT work occurs on a knowledge continuum: one where knowledge workers play a critical role at each stage [45]. In the software development arena, management practices applicable to American teams of companies may not be received well in other countries, so an understanding of cultural values involved is important as well [46]. Finally, Johnson cites similarities that exist between software and children's toys in that customer demand can change quickly and that these two items share short product lifecycles (among other similarities), and that risk involved in developing new toys can be mitigated through outsourcing [47]; the 24-Hour Knowledge Factory proposed here focuses on new software product development, and therefore a link to risk management and diversification obtained via this paradigm can be made as well.

3.3.4 Knowledge Management

Barthelmy observes that the management of the sourced effort is the major cost item, and that core knowledge should be kept within the core of the organization in order to manage this cost factor [48]. On the other hand, Lei and Slocum posit that sharing and understanding core competencies via alliances between sites is of utmost importance, and that it is imperative not to be "deskilled" [49]. The need to exploit knowledge networks the world over by utilizing capability based theory, where the process can act as leverage, is described by Tallman [50]. Powell states that the process of knowledge-creation, as well as the collaboration necessary to perform this function, is the primary core competency that a firm can exploit [51]. To further this observation, Davenport has shown that proper management of customer support knowledge is vital as this direct, human input is irreplaceable by automation [52]. The likelihood of success of companies depends on their ability to manage knowledge within their firms in order to ensure that duplicative effort is eliminated [53]. The 24-Hour Knowledge Factory model brings together those who work on similar tasks, furthering this ideal by ensuring that sharing of knowledge occurs between those who need to share it the most.

3.3.5 New Product Development

Researchers have identified several relevant factors in the creation of a successful global new product creation model. Coordination (as opposed to decentralization) between locations is critical, with firms that utilize multi-national new product development model, such as Boeing, being subject to different configurations [54]. Risks in new product development are always present, however; and the ones related to the loss of organization learning and innovations are highlighted by Earl [55]. Granstand et al. cite the need to both manage and diffuse technological competencies throughout distributed organizations, especially in new product development where different product groups may experience different results in terms of their technological competencies [56]. At Dell, for example, technology was utilized to change coordination activities between different areas of the supply chain, allowing all organizations involved to be treated as a one company [57]. This model suggests that vertical integrations can be altered to resemble more horizontal frameworks — while maintaining the vertical structure — solely through the introduction of sophisticated technology. This same concept drives the maintenance of globally distributed 'plants' that share tasks horizontally as proposed in the 24-Hour Knowledge Factory model.

3.4   24-Hour Knowledge Factory — Models and Hierarchy

The salient characteristics of the 24-Hour Knowledge Factory paradigm can be understood through a series of inter-related models. The first is a taxonomy for modeling tasks within the software domain. The next is a taxonomy for modeling organizational hierarchies that could be employed in building a 24-Hour Knowledge Factory. These axes provide the means for assessing the knowledge management needs of the particular application or operating environment, and also for managing the complex information flows within the organization. The information flow within a 24-Hour Knowledge Factory can be configured as a set of decisions — decisions regarding which projects to undertake, which people to consult on which projects, and which technical decisions to make with regard to all the stakeholders involved.

3.4.1   Taxonomy for Task Dependencies

The three scenarios depicted in Figure 3.2 illustrate the situations that may apply to a distributed software support center, a software maintenance engineering team, and a new product development environment respectively.

In the autonomous scenario (first case), individuals work in a relatively independent fashion, foregoing reliance on peer advice in the decision-making process. Software support centers - places where customers can phone in and receive knowledge from an individual support representative — are good

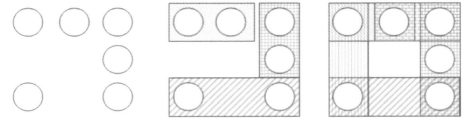

Fig. 3.2 Decision-Making Dependency Scenarios for Individual Work: (a) Autonomous — left, (b) Semi-autonomous — middle, (c) Tightly-interdependent — right.

examples of this type of work. In the semi-autonomous scenario (second case), individuals continue to work independent of their peers, but are given tasks that may necessitate consultation with others, such as more experienced peers, thereby creating a hierarchy. A software maintenance engineering team that develops incremental versions of an existing software product is a good example of this type of work — members of this team have the ability to work independently to a certain degree because code changes primarily come in the form of isolated bug fixes, but occasionally consult experts such as original code developers. The third scenario involves individuals who are heavily interdependent, such as members of software new product development team; input from other team members is consistently required, and decisions made by any one team member may impact the work of other team members.

3.4.2 Taxonomy for Organizational Hierarchies

Another axis to consider while considering appropriate decision support systems is the one for organizational makeup. In flat organizations, each and every decision-maker is, regardless of his or her geographic location or expected task, deemed part of a single organization. In other cases, hierarchical layers tend to appear within the overall organization, with hierarchical layers being added as work becomes more delineated. The three cases depicted in Figure 3.3 — based upon the degree in which task or geographical location is viewed in terms of importance — can be employed to filter decision making input and output through either geographic locations or task groups.

While flat organizations appear to be the simplest structure to deal with, the required interaction between individuals with no hierarchical oversight

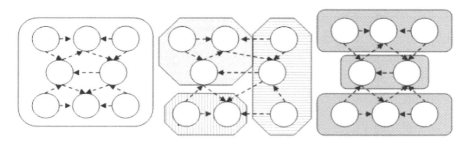

Fig. 3.3 Organizational models for heavily-interdependent decision making teams. (a) Flat organization — left, (b) Geography specific — middle, (c) Task specific — right.

function to aggregate and distinguish the myriad inputs and outputs of their decisions causes this structure to be the most complex in terms of the application of the 24-Hour Knowledge Factory concept. Consider a scenario in which software designers are located in US, Japan and Czechoslovakia, and testers are located in China and France. The stakeholders of the system may be located in other countries and time zones. Designers in US may need to consult testers in China for performance analysis, who may then need to consult stakeholders in Germany for performance measures. This complex decision-making scenario is more difficult to handle with a peer to peer approach in a flat organization model, as compared to organization models in which some notion of hierarchy has been pre-established, based on geographical or other considerations.

3.5   24-Hour Knowledge Factory — A Structure for Information Management

Almost all conceivable software development endeavors fit into the categories of taxonomies listed above. Some software endeavors are characterized be quick, coordinated decision making needs; in order for efficiency and effectiveness to be maintained, integrated decision support systems are necessary to properly mitigate the broad array of problems that are encountered when diffusing the decision making team around the globe. In the 24-Hour Knowledge Factory, work and knowledge are passed on to successive locations at set times; the knowledge transferred must be easily deciphered by the receiving team.

A prototype used to demonstrate this concept, called KNOWFACT (KNOWledge FACTory), was developed at MIT by two of the authors of this chapter. This prototype was tested in an actual satellite-design situation: technical decisions regarding specific system design were made on a daily basis, and required a constant re-evaluation process to analyze the effects these decisions had on a variety of geographically distributed stakeholders [58]. This type of model is very relevant for the programmers in order to understand requirements and to leverage past experiences and behavior to their advantage in future endeavors. The knowledge of historical issues and situations can greatly enhance the productivity of workers in the development, testing, and documentation phases in any 24-Hour Knowledge Factory.

The KNOWFACT decision support system as depicted in Figure 3.4. As shown, a Decision Rationale Module (DRM) facilitates the definition of key

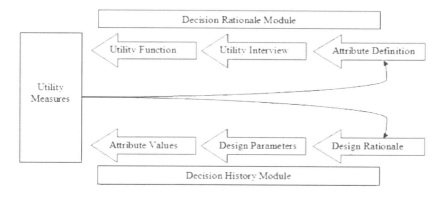

Fig. 3.4 Diagram of the KNOWFACT paradigm.

attributes that characterize the system for each stakeholder, while a Decision History Module (DHM) captures historical information on specific decision parameters. DRM attributes form the basis of a utility interview which then helps to determine the level to which the present state of the system satisfies the requirements of the stakeholders. The value of the DRM system, in terms of management decision making, lies in the structured and consolidated approach that it facilitates, partly by forcing the team to represent only the most important factors driving production of the software system. The values for the DHM parameters are aggregated to calculate values for the attributes defined by the DRM. This system was designed with earlier research in mind that showed the positive effect on human understanding and accuracy, in terms of interaction with decision support systems, when the system was decomposed into a small set of aggregate attributes; therefore, this system was developed to incorporate only a minimal set of aggregate attributes [59]. For best human interaction with the system, dynamic forms are utilized, as decision-makers have found decision support systems with flexible forms appropriate to the data being viewed to be the most useful [60]. Finally, the values calculated for each attribute are entered into the utility function to calculate an overall system utility measure; this utility measure is used by the team to redefine which attributes to use, and also to store the decision rationale on an evolving basis.

DHM provides the capability to perform automatic capture of vital information as the decision process evolves over time. It allows the user to view all the information in a graphical format, records the history of the states of the concerned variables, performs integrity checks and facilitates the capturing key elements. In addition, it provides a centralized repository of relevant information. DHM facilitates the creation of an evolving knowledge repository that contains knowledge on the current state of the activity being undertaken, the history of various states of the entities, and an automatic analysis of system integration integrity; the latter is performed with the objective of alerting the integrator about potential conflicts.

The DHM system is built to capture, reuse, and exploit information assets, all the while attempting to mitigate spatial and temporal barriers in large multi-disciplinary and multi-organizational endeavors. Applications from new product marketing campaigns to new system design are frequently re-conducted within organizations as new endeavors. Often, nominal amounts of knowledge, if any at all, are retained from previous processes or campaigns. A new method providing automated capture and storage of critical information (name, value, rationale, author, etc.) relating to each element of a project, along with a history of these data sets, was developed in an attempt to mitigate the problem of lack of reuse. Most of this functionality is automated, however users are asked to document rationale for specific actions at times of significant change to system design or implementation. A subsidiary and useful effect of this system is that it allows teams to reference decisions made on earlier projects that may have applicability to the decisions needed to be made now.

The DRM system is built to elicit and capture key information pertaining to the overall objective of the endeavor, and it analyzes and stores information about utility characteristics for each stakeholder involved in an endeavor. Prior research suggesting that cross-team decisions are best made when the decision can be characterized in terms of multi-attribute models and where each attribute represents an aggregated set of characteristics of the system were taken into consideration when drafting this function. The DRM system builds on this model by allowing stakeholders to interact at the attribute level, rather than having to use the attributes simply to break down the system into specific parameters. The particular parameters involved in a decision are related to the attributes once; then, utility measures are used to draw exact links between the parameter changes and the effect on the overall attributes.

The evaluation of the utility function, through the definition of attributes for every stakeholder, allows one to create, define and analyze attribute information, to structure the knowledge and to build a series of rule-based interviews with the objective of eliciting the stake holder's utility function. These interviews build upon earlier research on form structures to build the interview form. The interview questions are auto-generated based on existing templates and the data provided by the designers about the appropriate ranges for each attribute. These forms are dynamic, building on previous research that shows that dynamic forms lead to a higher quality of output. Additionally, the look and feel of the interview form has been designed to allow the users to visualize the utility implications of the input they are providing.

DRM facilitates the observation of links between multiple stakeholders, decision phases, and projects. It prescribes strict means in which trade space can be explored through the incorporation of preferences into decision criteria with methods based in economic and operations research theory [61]. Both cost-benefit analysis methods and utility analysis methods are utilized to obtain information from all stakeholders in a temporally and spatially decentralized design process, as well as to facilitate communication between those working in an organizationally decentralized environment. The principal mechanism for information acquisition is to conduct a series of interviews with stakeholders; these interviews are domain-specific and are designed to help capture users' preferences regarding the various attributes of the design architecture. These data drive the decision process via providing information about the utility and cost of each potential architectural alternative. In the 24-Hour Knowledge Factory environment, such a system would allow for the opinions of the decision-makers' to be adequately represented even when they are not available at the same time or same place.

3.6 Critical Success Factors for the 24-Hour Knowledge Factory

Diverse political, economic, cultural, and social factors would impact the efficacy of the 24-Hour Knowledge Factory approach for a particular software development endeavor. The list here is not exhaustive — instead, it can be utilized as the foundation for incorporating additional factors and dependencies. The relevant factors are as follows:

1. Location Choice
2. National Education System
3. National Labor Markets
4. Labor Costs

    5. Labor Language
    6. Cultural Differences
    7. Offshore/Onshore Labor Skills
    8. Location Time Zone
    9. Political Barriers
  10. Geopolitical Stability
  11. Demand Management
  12. Investment Diversification
  13. Communications Technology
  14. Internal Buy-In
  15. Organizational Model
  16. Long-Term productivity
  17. Dynamic Task Re-Allocation
  18. Offshore Value Chain
  19. Knowledge Product

The above list highlights the principal critical success factors, and relates to firms of all sizes — from startups and small enterprises to medium and large sized multinational firms — and in various global delivery models — from third-party vendor relationships to captive centers. Based on their broad relevance and significant potential impact, we discuss several of the important terms in the following paragraphs.

3.6.1   Demand Management

A 24×7 development model allows for much better management of customer demand. This model can promote a faster time to market for products, since companies can adapt more readily to potential changes in market conditions because of greater flexibility to reallocate and reassign resources, as well as to provide customers with access to skills they may not be competent in. A good example of the improved demand management is in the area of radiology, where offshore radiologists can read X-rays overnight and provide much better care, especially in an industry where the labor supply in the United States is limited [62]. As offshore knowledge workers gain experience and move up the learning curve and the value chain, this will lead to 24 hour availability of high value resources. In a company that employs home-based workers in India to perform medical transcription, the home-based work environment enables workers to be readily available at short notice. These workers can work longer hours as necessary, while continuing to manage family obligations at home, creating a new type of 'agile' knowledge workers operating in these knowledge factories in a real-time manner.

3.6.2 Long-Term Productivity

The use of the 24-Hour Knowledge Factory model facilitates access to more skilled labor for tasks previously performed only by lower skilled workers. For example: radiologists in the United States see reading X-ray results as a low-preference activity; radiologists in India, however, may see employment by a U.S. hospital as a high-value position without regard to the type of task being performed [12]. When moving toward a 24-Hour Knowledge Factory model, factors such as growth capabilities, quality management, and added communication and coordination costs accompanied by this type of paradigm must be incorporated into calculations of improved productivity [19, 22].

The key here is to "transform, not transfer" work [21]. The 24-Hour Knowledge Factory is a paradigm that involves transforming the tasks into ones that are more appropriate for the global model. As these tasks become transformed, jobs may be lost or redefined, and the important point in managing employees who are affected by these changes is to ensure that employees can continue to be productive and to add value. The "law of the horse" is an applicable analogy to this transformation: horse carriage manufacturers initially saw their jobs vanish after the automobile was invented; over time, the introduction of this 'disruptive technology' forced the horse carriage manufacturers to adapt [23].

Similarly, the advent of the 24-Hour Knowledge Factory paradigm will require pioneers adopting this paradigm to devote significant time in the beginning of the process in requirements gathering, organizing stakeholder workshops, and setting up communication norms, in order to realize major productivity benefits in the long-run.

3.6.3 Integrated Value Chain

The 24-Hour Knowledge Factory paradigm implies employing individuals at various locations; as different individuals move up the learning curve at various speeds, one can redistribute tasks appropriately between sites. According to Accenture, 51.9% of IT work offshored pertains to services such as maintaining computer networks; 36.7% pertains to solutions development such as website creation; and 11.4% pertains to leadership and management of projects [21]. There is a trend to move away from efficiency and towards growth, which can be seen through production activities being shifted from commodities to services to solutions, with vendors beginning to perform similar work for multiple customers [19, 21]. This movement up the value-chain of offshored workers coincidentally allows jobs remaining in a particular location to be more focused towards high-value work as well. One example is that of the U.S. doctor whose X-ray readings are done by a radiologist in India: the Indian doctor has moved up to a higher-value service, and the U.S. counterpart has freed up his or her time, and can consequentially focus on higher value-added tasks [20]. Current and future value-chain placement of knowledge workers of particular countries are influenced by geopolitical issues linked to these locations. China, for example, has lagged behind India in knowledge-based offshoring due primarily to a lack of English language ability amongst its citizens; as English becomes more commonly used in China. Russia saw many highly skilled PhD scientists and engineers experience dramatic drops in applicable tasks after the end of the Cold War; these domain experts are now making a transition into high-value services, such as optics design, at a lower cost to outsourcing firms [23]. The hiring process is a major factor in moving firms up the value chain, as a constant reevaluation is required of whether the foreign employees are indeed the highest skilled in the particular area of expertise [24].

3.6.4 Organizational Models

If flexibility is to be maintained, dynamic, evolving models that change with market conditions and take into account learning curve impacts on skill levels are needed. It is integral to properly assess the complexity of work required and to determine the specific skills sets available in each location when choosing a model, as matching to particular skills required is of great import [21]. A model may evolve in which similar functions or skills are located in multiple geographic locations; higher management overhead may result in trying to manage this unique situation, however greater returns may result as well, especially when taken in the context of the 24-Hour Knowledge Factory [24, 27]. Two matrices upon which the organizational models can be judged are coordination versus effort, and complexity versus project size11. These axes need to be frequently revisited, especially since engagements can commonly move from project-based engagements to long-term contracts that may require a different model [23].

3.6.5   Other Decision Factors

The 24-Hour Knowledge Factory embodies some of the secondary drivers such as quality improvements, improved access to expertise, and flexible staffing [24, 27]. In picking the right foreign partners for the 24-Hour collaborative endeavor, the final decision should be based on a thorough "kicking the tires" approach of visiting and testing the potential partner organizations and ensuring domain expertise, with an understanding that the lowest cost vendor may turn out to be the most expensive in the long-run if additional decision factors are considered [25].

3.6.6   Common Barriers within Firms

Typical firms exhibit significant barriers to employment of the 24-Hour Knowledge Factory; if this organizational structure is to be implemented, several of these barriers must be addressed during the initial phases of decision making in order to avoid having to take corrective action later on. Internal resistance — especially due to a perceived loss of control — may hinder proposed projects11. In addition, cultural, semantic and trust issues abound in these types of scenarios and must be analyzed at the beginning of an endeavor, recognizing the impact with respect to the interaction between knowledge workers in the 24-Hour Knowledge Factory [19, 24, 25, 26]. Even if the desire exists at all levels to pursue a globally distributed, collaborative endeavor, planning for process changes due to longer project planning cycles, more explicit definition of requirements and communication methods and the effects of ill-informed hiring decisions should be incorporated into the evaluation criterion [27].

3.6.7   Choice of Location

Inter-twined with the decisions related to whether to employ offshore resources, which model to pursue, and how to mitigate barriers, is the ultimate decision of which location or locations to pursue. Certain countries have identifying features which have made them popular locations for providing offshoring resources: India has a highly educated, English-speaking, IT skill base; China has an enormous labor pool [26]; and Russia has a legacy of USSR investments in science and a solid brand that has not been exploited to the level of India [23]. Factors to consider include the geopolitical stability of the country, the investment in education, labor and the skill set of a country's citizens, and the business environment in the country, including levels of corruption and the ease of setting up new businesses [19, 26].

3.7   Current 24-Hour Knowledge Factories — Case Description

Currently, only a handful of firms use the global work paradigm described in this chapter for purposes of new software development [63]. Some firms have adopted a similar, two-location version.

Gupta et al. [2] cite examples of both simple and complex software development projects involving multiple geographic locations [64]: one example involves developing a tactical, arms-length relationship with an Indian software vendor, with the benefit of being about to source relatively simple tasks only. In the context of our models discussed above, this example would suggest an "autonomous" task structure, with a "geographic" hierarchy, in which 24-hour development was not utilized, and thus the value of the Indian provider did not increase as time went by. Another example, cited in the same paper, involves a software team in Israel that serves as a strategic partner, accepting complex tasks and eventually sharing these between the home and sourced sites. This more strategic relationship allowed for the simultaneous utilization of skills available at both geographic locations, and tasks were completed in a more efficient fashion. This example, in terms of the three organizational models discussed earlier, conforms to the "heavily interdependent" task structure, with a "task-specific" hierarchy. Since emphasis was placed on the tasks to be performed, knowledge was transferred easily between the collaborating groups in the two countries, and progression of the relationship between these two sites saw ascension of the value-chain by the Israeli team in terms of the tasks being performed.

Ferdows cites organizations that have achieved much higher value from offshore teams in which they invested in a more strategic relationship [65].

Most strategic relationships involve greater sharing of knowledge. The real objective of the 24-Hour Knowledge Factory is to share tasks, along with the concerned knowledge.

One of the authors of this paper has worked extensively with the new software product development process of a large multinational firm that has a significant offshore presence. One division of the firm has tasks delineated into the "task-specific" hierarchy discussed earlier, assigning a team of between 6–15 persons to be in charge of specific product development tasks. These teams consist of members from disparate geographic locations; therefore, task sharing between multiple team members from different locations is a common practice. In interviews conducted with members of such teams, a variety of advantages and disadvantages of this work structure were identified [66]. These pros and cons are discussed in the following paragraphs.

3.7.1 Diversification of Knowledge Resources

The first step in bringing on new members into any team is to train these individuals by rapid transfer of knowledge. In the traditional offshore model, the completion of tasks results in knowledge that is then stored within the various global sites. In the autonomous task model, a specific piece of knowledge about a specific task is only held at the one location at which the task is completed, unless a method of knowledge dissemination is established. In the 24-Hour Knowledge Factory, with tasks being interdependent and shared between locations on a nightly basis, knowledge is disseminated as a natural part of the process, without any extra effort. The engineers interviewed cited this diversification of knowledge resources as being vital to their being able to assign any task to any location, on an as-needed basis. Thus, if a particular task had to be done quickly, a team manager could assign the task to a location that was just entering its daytime work hours. For instance, if a bug was found in software code at 5 pm U.S. time, the manager could assign to the task of fixing this bug to the Indian team without having to transfer significant amounts of knowledge to facilitate performance of this activity. Knowledge diversification to this level can be readily accomplished by global teams operating and sharing knowledge using the 24-Hour Knowledge Factory model.

3.7.2   Value Chain Movement

Daily communication between the onshore software team and the offshore software development team allows for the offshore teams to move up the value-chain much faster through this frequent communication — much faster than if a contract-vendor relationship was used. Interviews revealed a steady progression to higher-value tasks. Offshore team members were treated as new team members, and were inducted into the cycle through initial focus on simpler tasks such as fixing problems in the software. As these team members became more knowledgeable, they moved up the value-chain naturally, foregoing the need for a significant investment in training or knowledge dissemination.

3.7.3   Time for Resolution of Tasks

The interviews confirmed that tasks could be completed in a much shorter time frame once the offshore team attained a satisfactory level of knowledge. A one-on-one pairing of team members from each team to team members of the analogous, distributed team was adapted, allowing for limited investment in knowledge transfer at the beginning of a work day, as a consistent partnership supplemented understanding by knowledge-receiving workers. Thus, a task like fixing a bug, which might have taken 4 days to effect by one U.S. developer could now be done in half the time — only 2 days — by separating tasks and having a U.S. developer and an Indian developer work concurrently. Such an improvement in time-to-resolve was important as it greatly reduced the time to market for the product, and allowed new features to be incorporated quickly into a software product.

3.7.4   Earlier Reporting of Issues

One of the major uncertainties the development of a new software product is the untimely reporting of a bug late in the product cycle, which delays the release to market of the product. The software team which was interviewed used a testing team in China to work with developers in the U.S. to test pieces of the code overnight as the code was being developed. With the 24-hour model, a U.S. engineer was able to complete an incremental improvement to the piece of the code during the day, pass it to a Chinese test engineer with test instructions, and return the next day with results that helped to focus the efforts of the U.S. engineer and expose potential problems much earlier than if testing had been done later.

3.7.5 Unintended Process Improvements

The software team cited numerous process improvements brought about by the need to share information with other members of a geographically and temporally disparate team. Information such as design decisions, testing results, and review comments were tracked in databases. The transfer of knowledge was accomplished in an informal manner when team members were collocated, making the task of formal capture of knowledge significantly burdensome. The introduction of the 24-hour development model and the application of the decision rationale and history capture methods (e.g., KNOWFACT) became natural parts of the process, overcoming the natural tendency of software engineers to avoid or delay the task of dissemination of knowledge.

3.7.6 Cultural Understanding Incorporated within Software

Improvements made to the software through cultural knowledge integrated by each member regarding his or her culture and the usage patterns therein were noted. For instance, one Chinese tester pointed out that, due to a cultural issue that would prompt Chinese users to save face in front of their colleagues, Chinese users sitting in cubicles would not appreciate hearing a "beep" sound when they make an error while using software [67]. Another situation was described by an engineer in which a search function that allows for search by title rather than name was incorporated by Japanese developers when were considering requirements for selling into Japanese enterprises.

The 24-hour aspect of the development process is of great importance here because this allows for developers from different cultures to be engaged in each and every task, allowing for cultural input into all tasks at every stage of the process. Although software firms have many mechanisms for understanding requirements from different cultures, there is no substitute for the engineer building the system from the start with the cultural knowledge incorporated on a continuous basis.

3.7.7   Disadvantages of Decentralized Approach

A loss of informal communication was the most prominent disadvantage. Much of the software development process involves informal design meetings and reviews of design decisions. The engineers interviewed stated that they would often engage in informal discussions with fellow engineers; such discussions would lead to an unintended exposition of a major piece of knowledge that would have a significant impact on the overall project. It was often the case that engineers were not aware of the task history of their fellow engineers, and such informal communication would lead to a realization that the knowledge required to resolve a current issue could be provided by a fellow engineer. At the same time, the engineers interviewed consistently stated that the use of communication technologies between distributed teams can serve as a good replacement for informal interaction that occurs in case of collocated teams.

Overcoming this drawback, in the context of the taxonomies presented earlier in this chapter, would require a stronger understanding of the knowledge required in the information management framework and a better understanding of how tasks relate, in the context of the task dependency taxonomy. Once this understanding is developed, the hierarchy defined in the taxonomy for organizational hierarchies needs to facilitate the communication between the appropriate members of the process, so that the informal knowledge exchanges that are lost in the geographically distributed model can be recovered through deployment of alternative mechanisms geared to the characteristics of the particular organization.

3.8   Current 24-Hour Knowledge Factories in Action — Present State Analysis

The 24-Hour Knowledge Factory model is in limited use in the software industry today. As an example from industry, the mobile phone industry firm, WDS-Global, has utilized an 'Extreme Programming' methodology in a globally distributed, round-the-clock software development project [68]. The programming team was distributed across three sites (US, UK, and Asia) and each site had joint ownership of the code. The team was distributed to meet their customer's regional needs, while sharing the same code base to avoid redundancy and maintenance costs. The company hired coaches to teach, at two of the three sites, the practice of extreme programming, whereby programmers contribute to the same lines of code in tandem. The study found that there were slight cultural differences that caused some degree of confusion between the developers. Another problem they ran into was technical in nature — the amount of time it took to download code from certain locations slowed progress at times — increasing network bandwidth between the site in Singapore and the site in US and changing the technologies used to cache source control data were both cited as opportunities that led to improvement.

The WDSGlobal project demonstrated a set of enablers for the 24-Hour Knowledge Factory. The team met face to face at the outset to get to know and trust each other. Daily handoffs and explicit pairings within the team helped to retain and to enhance the trust relationship throughout the project. The team used VNC and video conferencing to "virtually" meet face to face and to share the work with each other. The daily knowledge transfer began as a summary of the day's work and evolved to discuss what the individuals had learned and objectives they had before they left. The project cited a key lesson learned as maintaining an equal sized team in each location, lest the design practice be monopolized by the location with the largest team. They also came to the realization that priorities change too quickly, so they determined that the managers should reprioritize tasks only once a week in order to ensure optimal team productivity.

Access and support for regional customers combined with a continuous engagement on the project served to make the WDSGlobal project a success.

3.9 Extending the 24-Hour Knowledge Factory Paradigm

A research team at the University of Arizona has designed CPro, a process addressing the operation of distributed teams within the 24-Hour Knowledge Factory environment. At its core is the Composite Persona (CP) [5, 6], representing a distributed team of developers that form a functional unit to which subcomponent development is assigned. Each subcomponent of a horizontally decomposed problem is therefore managed by a CP, whose members work on the subcomponent in series, with each successive shift continuing the work of the previous one. Multimind [5], a tool implementing the supporting elements of CPro has also been designed and implemented by members of this team.

As a functional unit, a CP has the properties of both an individual and a collective. In a global view, the CP has a unique name and acts as a singular entity between the disparate work sites. However, its collective nature allows the CP to act continually, regardless of the availability of individual sites. As the sites cycle between being active during the workday and inactive during others' workdays, the CP remains active, changing only the developer that is currently responsible for driving the CP's development. In this manner, development with CPs proceeds in much the same way as a traditional non-distributed process, although the individuals comprising the CP interact directly with each other to hold discussions and to resolve conflicts.

Software development processes are intended to improve the quality of the final product and to ease the communication between development stages for the participating developers. These processes are sets of rules and procedures that guide the knowledge flow as the project evolves through the development cycle. Processes with rigid rules and procedures, such as Tayloristic processes, relying on the separation of work into tasks assignable to specific functional roles, are high-ceremony processes. On the other hand, processes with few and flexible rules and procedures are generally considered low-ceremony and agile. Since a high-ceremony process will consume too much time for the efficient operation of a distributed CP, they are of little use here. Instead, low-ceremony rules and procedures make the Personal Software Process (PSP) [69] an attractive lightweight candidate for use. However, since PSP follows the conventional individual ownership of development artifacts, it is not directly applicable to the 24-Hour Knowledge Factory, but nonetheless it serves as the inspiration for CPro.

Multimind is an experimental tool developed at the University of Arizona with the aim of supporting the collaboration between CPs in a semantically rich environment [5, 6, 7, 8]. It aims to improve upon DICE and similar collaborative approaches to engineering. The foundation of Multimind is a Lifestream database, used to chronologically archive project artifacts in much the same fashion as a concurrent versioning system as well as to chronicle specific knowledge events. Examples of knowledge events that are logged into the Lifestream are developer actions such as reading or posting messages, executing searches, and visiting project-related web pages. This logging allows Multimind to correlate knowledge events and discrete evolutions of project artifacts.

Multimind, therefore, leverages existing resources and facilitates the creation of new artifacts in order to support both implicit and explicit communication requirements of the distributed CP. Explicit communication, where both the sender and the receiver of information intend to participate in communication, such as email or instant messaging, is a hindrance to distributed teams due to the spatial and temporal boundaries that define them. It can be argued that although the number of communication paths within a team increases as a quadratically with the addition of new team members, the amount of time spent on communication depends entirely on the form the communication takes. In Multimind, explicit communication is supported through the capture of text-based communications between developers in the context of the artifact under development. Implicit communication, on the other hand, is knowledge transfer that is based on communication through a shared mental-model of the task at hand. Expert teams rely on implicit communication more readily than novice teams, since the evolution of a development artifact is more easily conceptualized by experienced developers familiar with the artifact. Implicit communication is supported by Multimind through a facilitated visualization system, targeted to elevate the user's understanding of the software artifact by providing one or more visualizations that illustrate the coded logical structure, execution-related line-level statistics, or locations of execution errors within the executed code.

The operation of a distributed CP depends entirely on an efficient means of knowledge transfer, potentially realized by both reducing the necessity for time-consuming synchronizations between teams and by reducing the communication overhead incurred by the addition of new developers to a CP. As supported by Multimind, the combined facilitation of both explicit and implicit communication enables a comprehensive approach to coordinating the knowledge transfer both within and between the distributed teams comprising a 24-Hour Knowledge Factory.

Copies of Multimind code have been provided to counterpart universities in Poland and Australia, so that the efficacy of the overall approach can be evaluated in a real configuration with three collaborating centers in three different continents.

3.10 Future Applications of the 24-Hour Knowledge Factory — Breaking Barriers

Earlier in this chapter, the IBM case study was mentioned. This involved one-year long study of two teams: one with all members of the team located at a single location in the US, and the other involving a geographically distributed

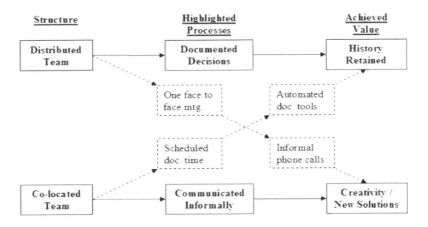

Fig. 3.5 Scenarios demonstrating how to leverage a 24-Hour Knowledge Factory.

configuration of the team. The detailed results from the controlled experiment can be readily extrapolated to provide valuable suggestions for teams that involve 3 or 4 different collaborating centers across the globe.

The geographic structure of the teams at IBM that were studied led to different forms of value being achieved from their knowledge-sharing processes. Figure 3.5 highlights that the structure of the distributed team led it to have a higher degree of documented decisions. Interviews with members of this team confirmed that one of the pieces of value obtained by this distributed process was that the history of decision making was better retained. Interviews with the co-located team indicate that it would not be feasible to enforce the same level of documentation on the collocated team; however this need should not be seen as a barrier to achieving the cited value of history retention. Instead, alternative processes such as scheduled time for documentation or the implementation of automated documentation tools could be used to achieve the same value for the collocated team. Similarly, the collocated team cited the informal communication as a process which led to higher degrees of creative and new solutions being found. Even though these informal meetings generally occurred face to face, the value achieved is something the distributed team can still obtain. Two suggestions provided in the interviews with the distributed team on this specific topic were a one-time face to face meeting that would introduce team members and bring a social component to the relationships, and the use of explicitly informal phone calls where no topic is predetermined so there is the potential to discuss any open ended topic.

Based in part on the IBM case study, the authors believe that the future of 24-hour knowledge factories is dependent on the reduction of barriers described in this paper, and a greater understanding of the potential benefits of such a model. For software firms to move towards this model, the following steps must be taken:

1. Assess the firm's software projects with respect to the task and hierarchy taxonomies presented above. By analyzing how interdependent the tasks and hierarchies are, one can determine how to move forward in developing a global delivery model that facilitates knowledge flow throughout the organization.

2. Assess the firm's decision rational and history system with respect to the information management framework. Determining whether the right information is properly acquired from the appropriate people and disseminated to the appropriate people — with as minimal burden as possible — will be vital to the successful operation of a 24-Hour Knowledge Factory.

3. Redefine the systems uncovered in Steps 1 and 2 to take advantage of the 24-hour offshore model. The decision to build a 24-Hour Knowledge Factory is not simply based on a current state analysis of the firm's systems, but a dynamic redefinition of systems to match the global delivery model which can yield the best results for the firm.

4. Work towards reducing the barriers exposed in the system dynamics model which may prevent the smooth flow of knowledge required for the 24-Hour Knowledge Factory. Such barriers may include communication technologies, economic and infrastructure support systems, training systems, organizational culture, and geopolitical risk management.

5. Build the 24-Hour Knowledge Factory! Note that even if the firm is not ready today to enter into the 24-hour global delivery system, it is important to begin the process of investigating and assessing the firm's own characteristics with respect to the taxonomies we have provided — the key point is that the 24-Hour Knowledge Factory

can be seen as an ideal endpoint on the spectrum of interdependent tasks and organizations, but firms can place themselves at any point on the spectrum and still succeed.

3.11   Conclusion

The 24-Hour Knowledge Factory model is an emerging model that can potentially bring the benefits of globalization to all parties involved. This chapter has focused on the modeling and implementation of the 24-Hour Knowledge Factory model. This model allows firms to integrate the contributions of key players from around the world and engage all contributors at all points on the value chain in the same tasks. Further, it provides a mechanism for engineers to maintain high-value input into tasks, while utilizing the cheaper labor of offshore partners to accomplish the lower-value tasks. This chapter used the example of the software industry as the pioneer in the emergence of this model, because the software process is based purely on the transfer and creation of knowledge with minimal infrastructure requirements.

The parallel nature of the data from the IBM case study — with positive findings on collaborative successes from each team — suggests that the common theme in the literature of geographic distribution being a barrier to overcome is not sufficient. Instead, geographic distribution should be seen as a potential asset that can be leveraged, along with time zone differences. A number of benefits from leveraging the geographic structure were cited in the interviews with the distributed team. Example include: an increase in documentation and history retention; the ability to share short term tasks which required immediate attention so that work could be performed around the clock; and a more structured definition of work tasks and distribution of work items. The methods of coding of archival data derived from e-mail, telephone, meeting and other interactions could also be used as a feedback tool that could be highly valuable for corporations seeking to quantify the impact of the 24-Hour Knowledge Factory. The evidence from this case study emphasizes that the spatial and temporal distributions in a 24-Hour Knowledge Factory should be looked upon as a characteristic to leverage rather than a barrier to overcome.

As this globally distributed work paradigm evolves, other industries will gradually adopt and benefit from the 24-Hour Knowledge Factory model.

References

[1] A. Gupta, S. Seshasai, S. Mukherji, and A. Ganguly, "Offshoring: The Transition from Economic Drivers Toward Strategic Global Partnership and 24-Hour Knowledge Factory", Journal of Electronic Commerce in Organizations, 5(2), 1–23, April–June 2007.

[2] A. Gupta and S. Seshasai, "24-Hour Knowledge Factory: Using Internet Technology to Leverage Spatial and Temporal separations", ACM Transactions on Internet Technology, 7(3), 1–22, August 2007.

[3] S. Seshasai and A. Gupta, "The Role of Information Resources in Enabling the 24-Hour Knowledge Factory," Information Resources Management Journal, 20(4), 105–127, October–December 2007.

[4] A. Gupta, Expanding the 24-Hour Workplace, Wall Street Journal, September 15, 2007.

[5] N. Denny, I. Crk, R. Sheshu, and A. Gupta, "Agile Software Processes for the 24-Hour Knowledge Factory Environment", Information Resources Management Journal, January–March 2008 issue.

[6] N. Denny, S. Mani, R. Sheshu, M. Swaminathan, J. Samdal, and A. Gupta, "Hybrid Offshoring: Composite Personae and Evolving Collaboration Technologies", Journal of Information Technology Research, January–March 2008 issue.

[7] I. Crk, D. Sorenson, A. Mitra, and A. Gupta, "Leveraging Knowledge Reuse and System Agiliy in the Outsourcing Era", Journal of Information Technology Research, April–June 2008 issue.

[8] K. O'Toole, S. Subramanian, and N. Denny, "Voice-Based Approach for Surmounting Spatial and Temporal Separations", Journal of Information Technology Research, April–June 2008 issue.

[9] C. Lovelock and G. Yip, "Developing global strategies for service businesses", California Management Review, 64–86, Winter 1996.

[10] A. MacCormack, L. Newman, and D. Rosenfeld, "The New Dynamics of Global Manufacturing Site Location", Sloan Management Review, 69–80, Summer 1994.

[11] A. Bartmess and K. Cerny, "Building Competitive Advantage Through a Global Network of Competencies", California Management Review, 78–103, Winter 1993.

[12] N. Venkatraman, "Beyond outsourcing: Managing IT Resources as a Value Chain", Sloan Management Review, 51–64, Spring 1997.

[13] M. Seitz and K. Peattie, "Meeting the Closed-loop Challenge", California Management Review, 74–89, Winter 2004.

[14] M. Blaxill and T. Hout, "The Fallacy of the Overhead Quick Fix", Harvard Business Review, 93–101, July–August 1991.

[15] G. Pisano and S. Wheelwright, "The New Logic of High Tech R&D", Harvard Business Review, 93–105, September-October 1995.

[16] V. Agrawal, D. Farrell, and J. Remes, Offshoring and Beyond, McKinsey Quarterly, No. 4, 2003.

[17] N. Kaka, A Choice of Models, McKinsey Quarterly. No. 4, 2003.

[18] C. Saunders, M. Gebelt, and Q. Hu, "Achieving Success in IT Outsourcing", California Management Review, 63–79, Winter 1997.

[19] J. Barney, "How a Firms Capabilities Affect Boundary Decisions", Sloan Management Review, 137–145, Spring 1999.

[20] N. Carr, "In-praise-of-walls", Sloan Management Review, 10–13, Spring 2004.

[21] C. Christensen, "The Past and Future of Competitive Advantage", Sloan Management Review, 105–109, Winter 2001.

[22] D. Light, "Cross-cultural Lessons in Leadership", Sloan Management Review, 5–6, Fall 2003.

[23] A. DiRomualdo and V. Gurbaxani, "Strategic Intent for IT Outsourcing", Sloan Management Review, 67–80, Summer 1998.

[24] J. Young, "Global Competition — The New Reality", California Management Review, 11–25, Spring 1985.

[25] A. Mizoras, In House versus Outsourced, IDC Opinion, IDC, 2004.

[26] J. Champy, "Is technology delivering on its Productivity Promise?" Financial Executive, 34–39, October 2003.

[27] F. McFarlan and R. Nolan. "How to Manage an IT Outsourcing Alliance", Sloan Management Review, 9–23, Winter 1995.

[28] M. Lacity, D. Willcocks, and D. Feeny, "IT Outsourcing: Maximize Flexibility and Control", Harvard Business Review, 84–93, May–June 1995.

[29] M. Lacity and R. Hirschheim, "The IS Outsourcing Bandwagon", Sloan Management Review, 73–86, Fall 1993.

[30] M. Lacity, D. Willcocks, and D. Feeny, "The Value of Selective IT Sourcing", Sloan Management Review, 13–25, Spring 1996.

[31] J. Quinn and S. Hilmer, "Strategic Outsourcing", Sloan Management Review, 43–55, Summer 1994.

[32] P. Fuchs, K. Mifflin, D. Miller, and J. Whitney, "Strategic Integration: Competing in the Age of Capabilities", California Management Review, 118–147, Spring 2000.

[33] D. Arnold and J. Quelch, "New Strategies in Emerging Markets", Sloan Management Review, 7–20, Fall 1998.

[34] G. Millman, "Going Beyond Commodity Outsourcing", Financial Executive, 55–57, September 2003.

[35] I. Cheifetz, "Think Like Buffet About How to Value Outsourcing", Financial Executive, 58, September 2003.

[36] H. Chesbrough and D. Teece, "Organizing for Innovation", Harvard Business Review. Best of Harvard Business Review, pp. 127–135, 1996.

[37] J. Quinn, "Outsourcing Innovation the New Engine of Growth", Sloan Management Review, pp 14–29, Summer 2000.

[38] M. Iansiti and R. Levien, "Strategy as Ecology", Harvard Business Review, 68–79, March 2004.

[39] T. Begley and D. Boyd, "The Need for a Corporate Global Mind-set", Sloan Management Review, 25–33, Winter 2003.

[40] G. Kumra and J. Sinha, The Next Hurdle for Indian IT, McKinsey Quarterly, 2003 Special Edition, No. 4.

[41] T. Kern, L. Willcocks, and E. van Heck, "The Winner's Curse in IT Outsourcing: Strategies for Avoiding Relational Trauma", California Management Review, pp. 47–70, January 1, 2002.

[42] M. Useem and J. Harder, "Leading Laterally in Company Outsourcing", Sloan Management Review, 25–36, Winter 2000.

[43] P. Taylor and P. Bain, "'An Assembly Line in the Head': Work and Employee Relations in the Call Centre", Industrial Relations Journal, 30(2), 1999.

[44] P. Drucker, "They're not Employees; They're People", Harvard Business Review, 70–77, February 2002.

[45] R. Aron and J. Singh, IT Enabled Strategic Outsourcing: Knowledge Intensive Firms, Information Work and the Extended Organizational Form, Knowledge@Wharton, The Wharton School, University of Pennsylvania, October 08, 2002.

[46] D. Elenkov, "Can American management practices work in Russia?" California Management Review, 133–157, Summer 1998.

[47] M. Johnson, "Learning from Toys", California Management Review, 43(3), 106–125, Spring 2001.

[48] J. Barhtelemy, "The Hidden Costs of it Outsourcing", Sloan Management Review, 60–69, Spring 2001.

[49] D. Lei and J. Slocum, "Global Strategy, Competence Building and Strategic Alliances", California Management Review, 81–97, Fall 1992.

[50] S. Tallman and K. Fladmoe-Lindquist, "Internationalization Globalization and Capability Based Strategy", California Management Review 116–136, Fall 2002.

[51] W. Powell, "Learning from Collaboration: Knowledge and Networks in the Biotechnology and Pharmaceutical Industries", California Management Review, 228–240, Spring 1998.

[52] T. Davenport and P. Klahr, "Managing Customer Support Knowledge", California Management Review, 195–208, Spring 1998.

[53] J. Quinn, "Strategic Outsourcing — Leveraging Knowledge Capabilities", Sloan Management Review, Summer 1999.

[54] M. Porter, "Changing Patterns of International Competition", California Management Review, 9–40, Winter 1986.

[55] M. Earl, "The Risks of Outsourcing IT", Sloan Management Review, 26–32, Spring 1996.

[56] O. Granstand, P. Patel, and K. Pavitt, "Multi-technology Corporations: Why They Have Distributed Rather than Distinctive Core Competencies", California Management Review, 8–25, Summer 1997.

[57] J. Magretta, "The Power of Virtual Integration: An Interview with Dell Computer's Michael Dell", Harvard Business Review, 73–84, March-April 1998.

[58] S. Seshasai and A. Gupta, "A Knowledge Based Approach to Engineering Design", AIAA Journal of Spacecrafts and Rockets, 41(1), January-February 2004.

[59] M. Bohanec and B. Zupan, "A Function-decomposition Method for Development of Hierarchical Multi-attribute Decision Models", Decision Support Systems, 36(3), 215–233, January 2004.

[60] J. Wu, H. Doong, C. Lee, T. Hsia, and T. Liang, "A Methodology for Designing form-based Decision Support Systems", Decision Support Systems, 36(3), 313–335, January 2004.

[61] R. De Neufville, Applied Systems Analysis: Engineering Planning and Technology Management, McGraw-Hill, New York, 1990.

[62] A. Gupta, R. K. Goyal, K. A. Joiner, and S. Saini, "Outsourcing in the Healthcare Industry: Information Technology, Intellectual Property, and Allied Aspects", Information Resources Management Journal (IRMJ), 21(1), 2007.

[63] R. Terdiman and A. Young, Trends in application outsourcing for 2003 and 2004, Gartner group, January 2003 (2003).

[64] A. Gupta, S. Seshasai, S. Mukherji, and A. R. Ganguly, Offshoring: The Transition From Economic Drivers Toward Strategic Global Partnership and 24-Hour Knowledge Factory, Journal of Electronic Commerce in Organizations (JECO), 5(2), 2007.

[65] K. Ferdows, "Making the Most of Foreign Factories", Harvard Business Review, March–April 1997.

[66] Interviews conducted by one of the authors of this chapter, in person, and via phone and electronic mail, with members of U.S. and Indian software development team, from April 9 to April 14, 2004.

[67] W. Hu and C. Grove, Encountering the Chinese: A Guide for Americans, Intercultural Press; 2nd ed. January 1999.

[68] M. Yap, Follow the sun: Distributed extreme programming development, Agile Conference, 2005 Proceedings, pp. 218–224, 2005.

[69] W. S. Humphrey, PSP: A Self-Improvement Process for Software Engineers, Addison Wesley Professional, March 2005.

4

24/7 Application in Medical Research

Martin Radlak

Wroclaw University of Technology, Wroclaw Poland
msc63mpr@cs.bham.ac.uk

Abstract

There is no doubt that cancer is very serious disease which every year affects millions of people all over the world. To reduce this amount, extensive work by research centers, pharmaceutical companies and charity organizations is conducted to develop methods of prediction, prevention and treatment. The purpose of this chapter is to introduce the concept of cancer detection using data obtained from SELDI-TOF-MS. This type of Mass Spectrometry has been proposed, because of its ability to produce high resolution spectrogram of proteins content in an organic sample. Assuming, that cancerous cells consist of proteins which are usually absent in healthy tissue, there is a hope to develop a method being able to distinguish between those two states, giving a solid base for real diagnosis which doctors have to make.

Keywords: Patient Diagnosis, Health Care Management, Mass Spectrometry, Mass Spectrometry Data Analysis, SELDI-TOF-MS

4.1   Introduction

The diagnosis of patients by health professionals is an important medical concern for modern health care management. Although, hardware is a breakthrough in the field it is still not precise enough to produce superior results expected from diagnosis equipment. Low repeatability of results, high noise and huge amount of data are only few of the difficulties encountered. Therefore, to supplement hardware deficiency, it is important to utilize more or less intelligent data analysis methods. Properly selected might, improve accuracy of the hardware. Here, the full process of detecting cancerous/healthy sample is presented: from raw data, through pre-processing to classification. Methods and algorithms, their properties and suggested implementation ideas are discussed. They are collected in an organized way and aim to present the state of the art over current research. Additionally, dependency of preprocessing methods into classifier performance is evaluated using, designed for the purpose of this project, analysis software. Whole research is described from a 24 h/day distributed work organization point of view. This form of research and projects realization, which includes many participants that are not limited by geographical location or time, may provide an excellent opportunity to accelerate current work in this and many other fields. Chapter is organized in the following way: at first, concept of Mass Spectrometry is briefly introduced. Secondly, methods used for data manipulation are described: from raw data derived from mass spectrometry, through preprocessing towards classification. Brief description of technology is presented. Next section is purely focused on organization of research and software development, thus main theme is application of 24 h work organization — introduced and analyzed in terms of this research.

4.2 Mass Spectrometry

Mass spectrometry is based on conversion from proteins contained in organic sample into ions, isolating them and detecting according to the ratio between mass and charge. This general definition introduce many technological problems with this approach of which the greatest one is ionization, i.e., how to ionize molecules to avoid splitting into smaller parts what could result in protein being detected as two smaller mass proteins.

According to [15], during the decade of the 1990s, changes in MS instrumentation and techniques revolutionized protein chemistry and fundamentally changed the analysis of proteins. These changes were catalyzed by two technical breakthroughs in the late 1980s: the development of the two ionization methods electrospray ionization (ESI) and matrix-assisted laser desorption/ionization (MALDI). But it is SELDI-TOF-MS technology, which is able to produce high resolution output systematically, even for large (easily breakable) particles.

4.2.1 Matrix Assisted Laser Desorption Ionization

MALDI creates ions by applying laser beam to a matrix which protects molecules from being destroyed by direct laser application. In contrast to older methods, MALDI allows ionization of more complex structures like peptides or proteins which are very fragile and loose its structure when ionized by traditional methods. In short, ionized particles are detected according to their mass to charge ratio. Each has the same charge applied, thus differentiate themselves according to their mass. Proteins are in gas state in the air. They are charged so applying electric field causes them to move towards opposite charge electrode. According to Newton's Law:

$$a = \frac{F}{m} \qquad (4.1)$$

For constant force F acceleration will depend on mass of molecule m. If the mass is bigger, acceleration is lower, such that smaller molecules will get faster to detector. By recording amounts of molecules arriving to detector surface, mass spectrogram can be produced, i.e., Figure 4.1. To summarize steps of proteins detection, which are performed by every mass spectrogram, Figure 4.2 presents these as a blocks diagram. Different technologies are better at some stages of this process, but still those steps are common:

- Sample introduction — the process of collecting tissue from probe into the system
- Ion source — mechanism to convert organic tissue into ions
- Ion analyzer — mechanism to organize samples in a way that detector will be able to distinguish differences between them
- Ions detector — process of converting ions into digital signal
- Data analysis — Hardware implemented procedures for data improvement

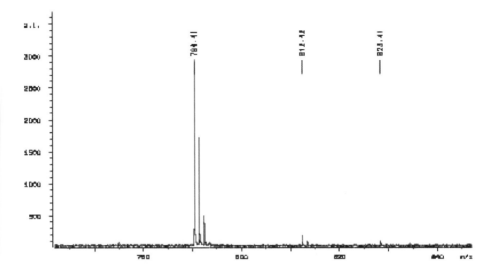

Fig. 4.1 Sample spectrogram.

Fig. 4.2 Spectrometry block diagram.

4.2.2   Surface-Enhanced Laser Desorption/Ionization

As every invention has to be improved, SELDI ionization method has been developed as an extension of MALDI. Major differences between those two methods have been described in application note of Ciphergen Protein Chip [13]. In both cases, proteins to be analyzed are co-crystallized with UV-absorbing compounds and vaporized by a pulsed-UV laser beam. Ionized proteins are then accelerated in an electric field, and the mass to charge ratios of the different protein ion species can be deduced from their velocity. The differences between SELDI and MALDI are in the construction of the sample targets, the design of the analyzer and the software tools used to interpret the acquired data.

In the SELDI method, protein solutions are applied to the spots of ProteinChip Arrays, which have been derivatized with planar chromatographic chemistries. The proteins actively interact with the chromatographic array surface, and become sequestered according to their surface interaction potential as well as separated from salts and other sample contaminants by subsequent on-spot washing with appropriate buffer solutions. The chromatographic surfaces provide a very good support for the co-crystallization of matrix and target proteins, resulting in the formation of a homogenous layer on the spot, thereby delivering an ideal crystalline surface for the subsequent analysis.

Sample preparation for SELDI experiments is quite different than the process for MALDI. For MALDI analysis, protein solutions are typically premixed with the matrix and dried on a passive surface. With the exception of flash washing with cold distilled water, on-target purification is not possible, and pre-target deposition sample cleanup procedures must be applied to reduce chemical noise and ion suppression. Also, on-target segregation of protein populations is not practical because the surface has only weak and unpredictable interaction properties. For these reasons, prefractionation using a variety of micro-techniques is often used. Taken together, these sample preparation requirements complicate the MALDI analysis, often resulting in sample loss as well as artificial qualitative and quantitative variances.

The analyzers used for SELDI and MALDI were designed with different purposes in mind. The ProteinChip Reader is especially adapted to achieve high-sensitivity quantification and good reproducibility. The ion source and detector are constructed to support very efficient ion transmission and ion detection over a wide mass range. The precise positioning of the laser beam is controlled by software both in manual and automatic mode. The process is visualized in a user-friendly format by a pixel raster map to facilitate the multiple analyses of the same sample spot, and software tools allow normalization of the resulting spectra to their total ion current for internal quantitative calibration. These features assure high precision and reproducibility even when great numbers of complex biological samples need to be comparatively analyzed. In contrast, MALDI devices are not designed for reliable quantitative precision over a wide mass range. They are a very good choice if high accuracy in the lower peptide range is needed without a requirement for high reproducibility of signal intensities. But if a good correlation between signal intensities and protein concentration is to be achieved over a wide mass and sample concentration range, the SELDI-TOF-MS-based ProteinChip Reader will always produce data with better reproducibility for hundreds of samples per day.

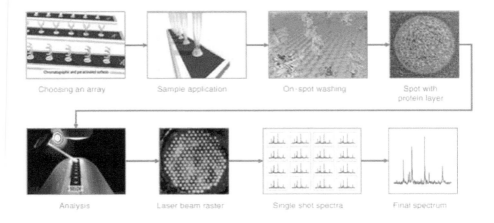

Fig. 4.3 Steps of sample processing for SELDI-TOF-MS.

4.2.3   Experimental Steps

According to [13] this process consists of following actions (also presented on Figure 4.3: after choosing an array from a selection of chromatographic and pre-activated ProteinChip Arrays, samples are applied and incubated on the spots. On-spot washing ensures efficient sample cleanup, and the spot surfaces allow the formation of a homogenous layer of co-crystallized proteins and matrix compounds. In the ProteinChip Reader, a laser beam is directed on the spot causing desorption and ionization of the proteins. A defined laser beam raster is used to selectively cover the entire spot surface and allows repeated reading of a single spot without using the same positions twice. Multiple spectra from a statistically meaningful area are then averaged in a final spectrum in which the mass-to-charge ratios of the ionized proteins are given and a good correlation between signal intensities and analyzed concentration is achieved for the different peptides and proteins in the sample.

4.2.4   Theory conclusion

After the technology process has been explained, it will be easier to understand what the actual problems are and limitations after the raw data have been produced. It is important to remember, that MALDI allows high accuracy to be achieved whereas SELDI-TOF-MS was developed to achieve better reproducibility over many experiments. Problems with reproducibility were described in detail by [4].

4.3 Sequence of Data Analysis

Previous section laid foundations of the technology used to produce raw data. In this kind of research idea is to be able to distinguish between normal or cancer disease. SELDI-TOF-MS produces data which in some way present structure of a sample. Intuitively, if a tissue is infected by a disease, it should contain different, specific to a disease, proteins. The problem is then to investigate samples collected and processed by mass spectrometer with respect to their spectrograms. Two randomly selected samples' spectra are presented on Figure 4.4. When figures are analyzed in detail, it is clearly noticeable that spectrograms for cancer (lighter line) and normal samples (darker line) differ significantly. This may confirm above claim, that it is indeed a good idea to process with research in this direction.

In the range 5790-5815Da there is undoubtedly significant difference between samples. By identifying similar areas — classification algorithm for detection would be straightforward to construct, based on the peak presence i.e. at 5800 Da.

At first glance, the task seems to be very straightforward, but in reality it is very difficult and there is no method which can produce 100% accuracy. Number of problems are encountered, which are the factors decreasing quality of the results.

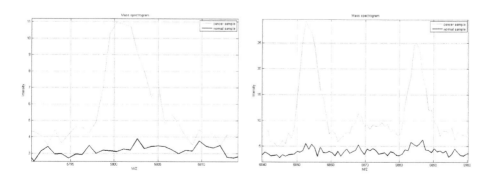

Fig. 4.4 Example differences between samples.

First of them is that it is very difficult to define which value defines major peak and which does only minor. Every sample is different and if a wider analysis is performed over many samples, conclusion is that the peaks are either much smaller or are not present at all in particular samples at the specific range. It is therefore very difficult to classify them. What makes task even harder is that generally, there are only few areas for which cancerous samples differ significantly from normal over all available input spectrograms.

There is also problem of overlapping samples. What is meant by this, is that when two samples belong to different classes (cancerous or healthy) but are not much different from each other, it is almost impossible to correctly classify them.

There are also many other difficulties when approaching this problem, i.e., presence of noise, shifts in values (vertical and horizontal), lack of normalization and many more with huge data size at the end. Finally, difficulties in developments and need for global cooperation within this field creates structural problem of managing and performing those kind of research.

Some of the methods, most commonly used, are presented in following sections. Many of them were implemented in the Mass Spectrometry Analysis Software (MSAS) designed in Matlab environment for the purpose of this project. This package has been chosen, because it is widely available through academic community. By developing software on a platform which is easily accessible, there is greater chance for further development and improvement.

Whole structure of the research in Mass Spectrometry can be organized into 3 main parts: preprocessing, analysis and classification, which are further divided into modules.

4.3.1   Preprocessing and Analysis

To be able to compare different samples to each other, it is important to prepare them for this process. Different factors have to be eliminated to prepare data for analysis and classification. If raw samples were not preprocessed, classifiers may detect normal sample as cancer, or cancer as normal what, if it was medical tool could result in very dramatic consequences. Person who is healthy could be mistakenly diagnosed as having cancer and would be directed for unnecessary treatment which is costly and very depressing. On the other side, ill person diagnosed as being healthy, may lose chance of being cured if the disease state is advanced. But preprocessing is not enough if wrongly applied. Therefore, analysis is a step to validate and define correct or wrong results. Most commonly used methods for these processes are: size and dimension reduction, baseline correction, normalization, peaks detection and alignment or noise removal. And this is the first major application for 24/7 research organization framework, which will be described in detail shortly.

4.3.2 Classification

Classification, on the other hand, is a supervised learning algorithm. What this term means is that algorithm is trained to classify data into predefined classes. Performance of classification is measured as percentage of correctly/negatively classified samples. It is straightforward task to test the performance, because for each input, in training set, specific output is assigned. The goal of classifying algorithms is to minimize error between desired and actual output.

There are many factors which influence how classifying algorithm performs. One of the reasons of poor classification is when data are overlapping — classes are not easily separable. This is caused when data sets have similar parameters, so it cannot be straightforward to assign to one of the classes. Problem is presented on Figure 4.5. Therefore it is important to use a classifier, which will be able to extract hidden information in data set, distinguish differences and classify samples correctly.

As suggested in [16], that classification algorithms can be divided into two major groups: those algorithms which depend on the data, initial conditions and experiment conditions on its own. They tend to return different results for each iteration. With those algorithms, often random population of solutions is initialized and fitness of these solutions is improved for number of generations, i.e., Evolutionary Computations or Neural Networks. Initial conditions are often the reason why method performs better during one run and worse during another. These methods are called heuristic.

Second group of algorithms are strict mathematical models. Those algorithms always follow determined steps thus for specific input, always return the same output, even if the initial conditions are different each time. These methods are called deterministic or exact models, i.e., PCA, LDA, k-Nearest Neighbors.

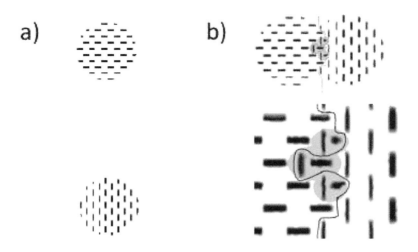

Fig. 4.5 Classes: (a) Well separated, (b) Overlapping.

An overview of the methods, which can be applied to mass spectrometry data has also been described by [20], thus it will not be discussed here.

4.3.3   Verification

Every classifier has to be verified to show how well does it perform for input data. In this case data set is divided into 2 subsets: training and verification set and it is important to perform this process properly. For mass spectrometry data freely available on the Internet, number of samples is limited thus algorithms have been proposed to increase the size of it. But the easiest ways is to permutate data set and choose desired number of randomly selected samples for training and use remaining set for testing. Another methods used for the small data set is boosting or bagging [7].

When classifying system is ready — it is important to perform statistical tests to evaluate its performance. It is crucial to test it on a data which have not been used for training. Performance should be tested for some number, i.e., 50 independent runs and mean and standard deviation should be calculated. This will create statistically significant sample and if mean, according to t-test is significant, there is high probability that this is indeed true performance of the classifier.

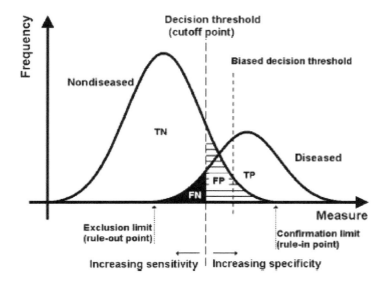

Fig. 4.6 Probability density of two class samples (After: [8]).

In general, performance can be expressed as a percentage of correctly classified samples. But this is not enough. Often, classifier may perform very well detecting existence of disease, but poorly when rejecting its existence. Other classification algorithm may perform well when rejecting existence of disease but poorly detecting its existence. This is strictly due to type I or type II errors. In medicine, more harmful would be one, which under performs while detecting diseased samples. The reason is because it is better to false predict that the person is diseased, because then further investigation can be conducted. If, on the other side classifier would classify diseased person as healthy, and no further investigation would be conducted, person would lose its early chance to be cured. And time is the key factor when it comes to cancer disease. If diagnosed at early stage, chances for person to be cured are very high.

Therefore, instead of measuring only percentage of correctly classified samples, two other values are also examined. They are called PPV[1] and NPV.[2]

According to 3.6 they are calculated as follows:

---

[1] Positive Predictive Value
[2] Negative Predictive Value

1. PPV — indicate the percentage that in case of a positive test, patient indeed has the specified disease;

$$PPV = \frac{true\ positive}{true\ positive + false\ negative} = \frac{TP}{TP + FN}$$

2. NPV — indicate the percentage that in case of negative test, patient indeed is healthy;

$$NPV = \frac{true\ negative}{true\ negative + false\ positive} = \frac{TN}{TN + FP}$$

Further details about validation requirements of classifier were described in work by [8].

4.4   Software Design Elements

Up until this moment, there was no thought about 24/7 work organization. This has been done deliberately, in order to lay solid foundation of the core problems in described research type. Here, whole problem will be presented from a perspective of 24/7 because it may significantly increase speed of research for diagnostic software, which has always been in very high demand and together with increasing standard of living; the need for it is growing more rapidly. 24/7 system is an approach to software development which is based on cooperation of few research groups located in different time zones.

4.4.1   The Need for Distributed Project Realization

There is no doubt that researchers, project, managers, programmers — people work at their best during daytime. It is very well known property of human brain to be active during the day and regenerate over night. Therefore, a single group working on a project performs it effectively only through 8 hours a day and sometimes even less. Thus if task requires 320 hours — it would take 40 days to accomplish it. No speed up is possible because of two major factors:

1. Workers would have to extend their working day
2. More people would have to be employed.

First solution would only "virtually" boost the speed of a project, because participants would have to work when their brain activity is lower, thus productivity is reduced. This solution is not recommended, although used for some periods of time within companies or other groups working on the same task, i.e., when a deadline approaches.

Second solution might be difficult to evaluate, especially if work is conducted in a very narrow area, where supply of specialists is very limited. It may take a long time to recruit required amount of people from within reasonable area. Additionally, this process requires further financial expenses which are not the case here. Therefore, 24/7 work organization might be highly applicable.

Detailed description, advantages and disadvantages 24/7 is a solution for described problems regarding traditional process of software development. It is a framework of effective project management which is intended to significantly speed up the software development. It is based on few research groups cooperating together, for ease of explanation, 3 different research centers. The key idea is that they are located in different time zones all over the world, i.e. Poland, United States and Australia as proposed by [23]. This situation is illustrated on Figure 4.7.

Preferably, as Figure 4.7 presents, those research centers should be placed within reasonable time zones, not just at any randomly chosen location. Although it is possible to locate centers anywhere in the world, systematic placement is required to fulfill the aim, that only one group works at its highest

Fig. 4.7 Time zones and example research centers arrangement.

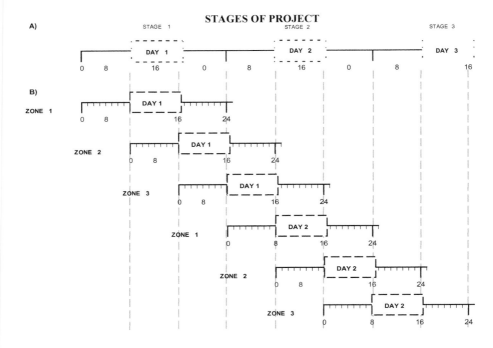

Fig. 4.8 Schedule of work: (a) one research center; (b) 3 research centers.

activity time (during regular working hours). None of the other groups work simultaneously on the same task, during the same time although marginally, they can overlap, i.e., to exchange information between themselves. After first group working day is finished, results are passed to the subsequent team, located in different time zone (i.e. 8 hours difference) where the working day has just begun. Process is repeated further as long as the project duration. To visualize the concept just described, Figure 4.8 has been provided.

In case of research studied here, there are tremendous possibilities available for application of 24/7. At first, whole research can be easily modularized. Each part of the process consists of methods, out of which there is a choice, whether to apply a method or not. Additionally, even if method has already been developed, during the process of testing, researchers may identify which of the algorithms perform poorly, i.e., lead towards reduction of performance. Then, a task might be to improve this specific algorithm. It would not block the process of testing and identifying other poorly performing modules, as the whole process can be run with any combination. Due to this property, through extensive experimentation robust system can be constructed.

4.4.2 Advantages and Disadvantages

Advantages of proposed solution are clearly presented in Figure 4.8, if 48 hours were considered. By the time single research center finish their second stage of the project, 3 research centers will accomplish 6 stages. This is the most significant advantage of 24/7 work organization. Projects could be realized almost three times faster. The work "almost" means that it is not exactly 3 times faster due to additional communication processes which has to be performed when using 24/7 system. One of them is synchronization of the work between the changes of shifts. It may take another several minutes for a group which intercepts results of previous group to fully understand what has been done within last few hours. Another difficulty with 24/7 approach is task sequencing. It is similar to a pipeline processing, but a group is not tightened to any goal that have to be achieved. If they do not finish what was assigned to them, they simply explain following group what was the difficulty and the work is carried on. This situation may even be with advantage for the project: when one cannot make a progress with a difficult part, fresh look of the following group may provide some new ideas and move the project forward — probably further increasing the speed of works. Finally, this kind of work organization may introduce kind of healthy competition when each of the group would like to progress more than the others.

By knowing the framework, following sections will propose ideas of employing 24/7 into Mass Spectrometry research as it provides opportunities for this to be used, but first of all, key factors of software design will be described.

4.4.3 Steps of Software Design

Most of researchers who work in the field of Proteomics come from Computer Science, Mathematics or Biology background. They develop and improve algorithms which in future may be applied for clinical diagnosis. By having knowledge of software programming, it is common that designed programs are difficult to use for someone without those skills. This is the reason why it is very important to consider requirements of end user, because it is the person who will be using software the most. In general, it will be addressed to doctors with no high proficiency in programming. Friendly interface must comply with following rules: simple, intuitive, nice presentation, ergonomic — and finally, should make process of analysis easier. Some very useful tips on GUI design were explained by [19].

- **Analysis:** Designing a software is very laborious task. It can be even more time consuming if there is no planning included. That is why very important aspect is to spend more time on the process which will allow very good preparation. Analysis should include the following steps, and if performed properly can save huge amount of time:

    1. Decide who will be the end user of software

    2. Identify how the software is going to be used

    3. Be aware of computer limitations

    4. Plan the software in a way that it will be easy to upgrade

    Above list contains very important steps and in case of 24/7 framework, this should be performed by each group separately. Then a meeting of all members would be desired in order to exchange ideas and construct final plan for the project, where each of possible obstacles would be described and planned to avoid.

- **Design:** With the first step, analysis properly performed, many important aspects should be noticed which will help to avoid faults during design process. When designing, there are also many subcomponents of the process, which must be considered. First of all, it is worth remembering that main goals of GUI are usefulness, reliability and that it should make work easier. As the core application of software being designed for this project is data processing, or data mining in other words — GUI should provide great flexibility. It should not overwhelm the user and avoid states where user does not know what is happening, i.e., when the function, which requires long time to accomplish, is launched, user should be informed how much more time will it take (in this case progress bars should be considered). Analysis software should reduce the demand on the user so he can focus on performed experiment rather than on technical issues of programming.

- **Paper prototyping:** After an effort has been made to analyze the requirements, with conclusions after design process; it is a moment to prototype an interface on paper. This very fast technique will give a brief overview of the ideas, how will the software look like on the screen. It is much better to see how everything looks like, before it is programmed. Changing layout on a paper is also much easier and faster. It is more efficient to program correctly from beginning rather than having to reprogram parts of the software. This step of the process, if performed properly, will save huge amount of time in the future.
- **Construction:** When all the previous steps have been executed with success, the GUI can be programmed. It will be much easier, time efficient and will result in the software which is useful, and which will help to perform tasks it is designed for.

4.4.4 Implementation of Proposed Solutions

With 24/7 work organization scheme, jobs have to be scheduled. In traditional distributed task assignment process, everything had to be planned carefully. If tasks were partitioned improperly, some groups were inactive, due to delay of a team which results they were dependant on. Usually, to manage many groups, one centralized unit had to be created. Exchange of information was a long process, thus gain achieved by distributing tasks was much smaller than expected. 24/7 system helps to avoid many disadvantages of traditional distributed work scheduling.

With Mass Spectrometry research, it is possible to make advantage of 24/7. As described earlier — Steps of data manipulation — it is clear that this process is very structured and can be simply divided into modules, i.e., data preprocessing, data manipulation or classification. Those are only 3 general steps of data manipulation, but can be further divided into smaller modules. This is excellent property to apply distributed scheduling. Traditional distributed tasks assignment could be one choice: each group would have to work on different module. But to avoid disadvantages of traditional scheme, 24/7 is proposed to be used. Jobs can be assigned much more loosely so there is no requirement that a group has to finish day goals to not to block another group. It is based on great cooperation between teams, which may encourage new ideas.

4.5   Case Study

The main goal of this work was to prepare a mass spectrometry data analysis software for data classification. The goal has been achieved — working program has been written and used for testing preprocessing and classifying algorithms. Some of the ways of using the software will be presented as follows.

4.5.1   Data Set Characteristics

Example data set was obtained from National Cancer Institute.[3] There are different sets to choose from, but the most recent one has been chosen: High Resolution SELDI-TOF Study Set. It contains 216 samples out of which 121 were derived from population with ovarian cancer, 95 were derived from healthy population. This data set has not been preprocessed so it is a good input to test preprocessing methods and classifier. More on the data set has been written by [2] and [3].

4.5.2   Program Requirements

Mass Spectrometry Analysis Program has been developed using Matlab 7.1 environment. It requires standard Matlab 7, Bioinformatics Toolbox and Statistics Toolbox, which are supplied with standard edition. It does not require any other software, but the better hardware used, the faster and more pleasant will be the work with program. The most computing power consuming process is loading data which are stored in separate text file for every sample i.e., it takes approximately 10 minutes to load all samples on the Intel Centrino 1.5 GHz with 1024 MB of RAM memory. There are no significant delays on any of the functions which were used in program so Pentium IV 1.4 GHz processor with 1024 MB of memory might be used as a benchmark.

---

[3] http://home.ccr.cancer.gov/ncifdaproteomics/OvarianCD_PostQAQC.zip

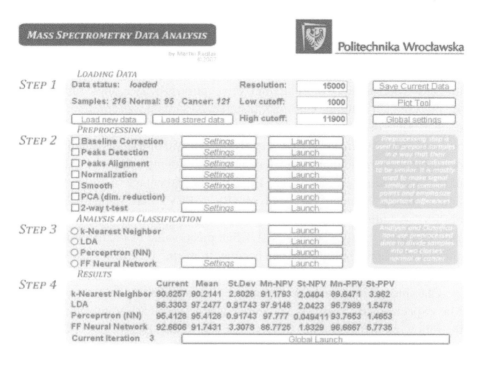

Fig. 4.9 Screenshot of the mass spectrometry data analysis program.

4.5.3 Program Demonstration

When the program is launched, it looks as presented on Figure 4.9. Core structure of the software is intuitive and follows the sequence of a trial in 4 steps:

1. Step 1 — Loading data;
2. Step 2 — Preprocessing;
3. Step 3 — Analysis and classification and finally;
4. Step 4 — Results.

On the right hand side there are few options which allow to perform additional tasks. Those are: Saving current data (this can be done at any time), plotting current data and setting global parameters.

4.5.4   Results

Results are presented according to which classifying method has been used. It is a percentage of correctly classified samples. There are few different possibilities to see performance of a classifier. First option is to try a single run. Results are presented in column "Current". If statistical sample was required, it can be done using "Global Launch". This evaluates selected preprocessing methods and performs classification for every classifier — 50 times. Mean and standard deviation is then calculated and displayed in following columns: "Percentage of correctly classified", "Standard Deviation of percentage of correctly classified", "NPV", "Standard Deviation of NPV", "PPV" and "Standard Deviation of PPV". Additionally, to test for significance of the results — Student's T-test has been used. Null hypothesis H0 is that mean value of results without any additional preprocessing methods is equal to mean value of results with particular preprocessing methods applied.

Classifier performance according to preprocessing methods. In this part Mass Spectrometry Analysis Software has been used to examine efficiency of the following methods: k-Nearest Neighbors (KNN), PCA+LDA (PCA), Perceptron (PER) and Feed-Forward Neural Network (FFN) trained with Back-Propagation algorithm. Therefore, dependency of preprocessing methods in final classifier performance has been examined. Sample data has been equally divided into training and evaluation set. Training and classification has been repeated 50 times to create statistically significant results. Training and evaluation data were randomly selected for each run to avoid repetition of the same set. Results are presented in Table 4.1. It is not straightforward to say that applying preprocessing methods will increase performance of a classifier. Some of the methods decrease its performance, i.e., when only normalization has been applied; performance of LDA has been increased. Also, standard deviation has been lowered indicating that classifier produce much more repeatable output. On the other side, when three different preprocessing methods were applied together — Baseline Correction, Normalization and Smoothing — apart from k-Nearest Neighbors, all classifiers performed worse than if no preprocessing was applied. This leads to following conclusion: at first — preprocessing methods should be carefully chosen to not loose important information which raw data contain, secondly — performance of the classifier might be increased

Table 4.1 Classifier performance based on preprocessing method.

	KNN	LDA	PER	FFN
No preprocessing				
% Correct	91.74	98.28	96.44	92.33
St. Dev.	3	1.12	1.8	4.4
NPV (Mean)	89.24	98.66	98.02	93.32
NPV (StDev)	4.75	1.84	2.67	7.18
PPV(Mean)	92.06	98.07	95.46	92.79
PPV(StDev)	3.66	1.9	2.82	5.76
Baseline Corrected				
% Correct	91.25	98.35	96.5	93.19
St. Dev.	3.16	1.2	1.78	3.5
T-test	0	0	0	0
NPV (Mean)	90.83	98.41	97.84	95.9.48
NPV (StDev)	5.12	1.88	2.35	4.93
PPV(Mean)	91.89	98.37	95.68	92.22
PPV(StDev)	3.24	1.72	2.94	6
Normalization				
% Correct	90.81	98.68	97.27	94.24
St. Dev.	2.97	0.89	2.01	4.24
T-test	0	0	0	0
NPV (Mean)	89.08	99.1	98.2	94.47
NPV (StDev)	5.08	1.43	2.39	7.71
PPV(Mean)	92.6	98.49	96.74	95.01
PPV(StDev)	3.25	1.49	3.11	4.1
Smoothing				
% Correct	90.81	97.96	95.65	93.28
St. Dev.	2.94	1.4	2.14	3.63
T-test	0	0	0	0
NPV (Mean)	88.87	98.62	97.33	94.81
NPV (StDev)	4.74	1.42	2.93	5.78
PPV(Mean)	92.71	97.53	94.66	93.02
PPV(StDev)	3.01	2.2	3.27	5.29
Baseline Correction + Normalization + Smoothing				
% Correct	92.09	98.28	95.95	91.27
St. Dev.	2.94	1.17	2.39	13.67
T-test	0	0	0	0
NPV (Mean)	91.19	98.11	94.66	88.17
NPV (StDev)	4.75	2.27	4.4	14.33
PPV(Mean)	93.12	98.49	97.28	94.82
PPV(StDev)	3.24	1.57	2.48	14.17

by applying correct preprocessing methods, but it is important to notice, that if the classifier is well constructed, it can be superior, even if no preprocessing is applied (i.e., LDA + PCA).

At this point it is also important to comment on results achieved by using Feed-Forward Neural Network. As observed in Table 4.1 FFN performs poorly. It is better than KNN but is characterized by highest standard deviation of the results. The reason for it is probably the way of using Neural Network in this work. Tests were performed for 3 fully connected layers consisting of 9-5-2 hidden units. Network has been trained using simple back-propagation algorithm, even though the architecture of the network has not been set to analyze data specifically, results produced were satisfying. Therefore, there is a chance to achieve increased performance, if a better Neural Network model has been prepared, i.e., different activation functions, architecture optimized using Evolutionary Computation or more stable training algorithm, but this has not been tested here.

Finally, there is a very important conclusion to draw from obtained results: all of the results are significantly different than default settings (when no preprocessing has been applied). This is strong evidence, that it is very difficult task to increase performance of classifier by few points. Therefore, there is still a lot of work to be performed in order to develop even better algorithms.

4.6   Results and Conclusion

Methods which allow noninvasive or low invasive diagnosis play very important role in present medicine. They show the state of a patient without making him feel any discomfort while extracting data. These methods allowed better diagnosis made by doctors thus survivability and comfort of population has been increased recently. This is the main reason why so much research is done to develop new methods which will allow examining patient non-invasively and further increasing a comfort of prevention or treatment process [8]. In his article described the requirements of diagnostic application. He mentions three most important expectations of this kind of tool which are:

1. It is essential to provide diagnostic tests that allow for definite and reliable diagnosis tied to a decision on intervention (prevention, treatment or non-treatment)
2. It is essential to meet stringent performance characteristics for each analysis (in particular: test accuracy, including both precision of measurement and trueness of the measurement)

3. Provide adequate diagnostic accuracy (i.e., diagnostic sensitivity and diagnostic specificity, determined by the desired positive and negative predictive values which depend on disease frequency.

SELDI-TOF-MS combined with appropriate data analysis software could, as presented in this chapter, possibly be a tool to meet desired requirements. But there is still a lot of work that needs to be done in order to achieve satisfactory results. But as there are many tools available, low mass spectrometry reproducibility (which is the biggest problem at the moment) can be supplemented by good analysis software which will be able to correct deficiency of the hardware. This is why this work has been conducted: to make a step towards better analysis software. Conclusions drawn from the results MSAS provides may be very helpful for future developments of similar software. It is therefore worth putting an effort into this field of research, and methods should be developed to speed up development of this research. One idea is to encourage research centers all over the world to group into international teams, working towards the same goal. This is the fundamental aim of 24/7 — round the clock work organization.

As a result of described work, Data Analysis Software for Mass Spectrometry has been developed. Main goal set for this project was to develop data analysis software and to analyze its performance — this has been achieved. There are many strengths of this development. Firstly, the software is based on Matlab environment which is a cross field development platform. Not only programmers use it, but also biologists, chemists or physicists. By encouraging these groups to create communities which will cooperate using, i.e., proposed 24/7 system. Researchers from different areas of science should be able to communicate in the same, well known programming (scripting) language. This is required, as i.e., biologist know details which computer scientists know nothing about (i.e., structure of organic tissue — biologist, programming skills — computer scientists).

Additionally, many algorithms and general framework for cancer detection have been presented. This allowed lying solid foundations for future work which should be directed towards developing an online application: accessible by anyone — everywhere. This online system could possibly collect data about patients, i.e., their medical history, which would provide additional knowledge. Furthermore, based on these data, new data mining algorithms could be employed to analyze this information and possibly find the causes of cancer disease.

Acknowledgements

This project introduced mechanisms and ideas which are not sole work of the author. Information provided here are parts of global research community, where there is a growing need for synchronization of work they perform. By introducing author's point of view, possibly new ideas will emerge through further researchers. Work has been funded by EPSRC scholarship.

References

[1] Oxford English Dictionary, Oxford University Press, 2006.

[2] T. P. Conrads, V. A. Fusaro, S. Ross, D. Johann, V. Rajapakse, B. A. Hitt, S. M. Steinberg, E. C. Kohn, D. A. Fishman, G. Whiteley, J. C. Barrett, L. A. Liotta, E. F. Petricoin III, and T. D. Veenstra, "High-resolution Serum Proteomic Features for Ovarian Cancer Detection", Endocrine-Related Cancer Journal, vol. 11, 2004.

[3] K. A. Baggerly, S. R. Edmonson, J. S. Morris, and K. R. Coombes, "Emphhigh-resolution Serum Proteomic Patterns for Ovarian Cancer Detection", Endocrine-Related Cancer Journal, vol. 11, 2004.

[4] K. A. Baggerly, J. S. Morris, and K. R. Combes, "Reproducibility of SELDI-TOF Protein Patterns in Serum: Comparing datasets from Different Experiments", Journal of Bioinformatics, 20(5), 2004.

[5] Zhi-Hua Zhou, Senior Member IEEE, Xu-Ying Liu, Training Cost-Sensitive Neural Networks with Methods Addressing the Class Imbalance Problem, IEEE Transactions on Knowledge and Data Engineering, 2005.

[6] Jiangseng Yu and Xue-Wen Chen, "Bayesian neural network approaches to ovarian cancer identification from high-resolution mass spectrometry data", Journal of Bioinformatics, vol. 1, 2005.

[7] David Donald, Tim Hancock, Danny Coomans, and Yvette Everingham, Bagged Super Wavelet reduction for boosted Prostate cancer classification of SELDI-TOF mass spectral serum profiles, Chemometries and Intelligent Laboratory Systems, vol. 82, 2005.

[8] F. Vitzthum, F. Behrens, A. N. Anderson, and J. H. Shaw, "Proteomics: From Basic Research to Diagnostic Application A Review of Requirements and Needs", Journal of Proteome, vol. 4, 2005.

[9] J. S. Yu, S. Ongarello, R. Fiedler, X. W. Chen, G. Toffolo, C. Cobelli, and Z. Trajanoski, "Ovarian Cancer Identification Based on Dimensionality Reduction for High-throughput Mass Spectrometry Data", Journal of Bioinformatics, 21(10), 2005.

[10] Lindsay I Smith, A Tutorial on Principal Components Analysis, 2002.

[11] G. A. Satten, S. Datta, H. Moura, A. R. Woolfitt, Maria da G. Carvalho, G. M. Carlone, B. K. De, A. Pavlopoulos, and J. R. Barr, "Standardization and Denoising Algorithms for Mass Spectra to Classify Whole-organism Bacterial Specimens", Journal of Bioinformatics, 20(17), 2004.

[12] K. R. Coombes, S. Tsavachidis, J. S. Morris, K. A. Baggerly, M.-C. Hung, and H. M. Kuerer, "Improved Peak Detection and Quantification of Mass Spectrometry Data Acquired from Surface-enhanced Laser Desorption and Ionization by de-noising Spectra with the Undecimated Discrete Wavelet Transform", Journal of Proteomics, 5(16), 2005.

[13] S. Vorderwülbecke, S. Cleverley, S. R Weinberger, and A. Wiesner, "Protein Quantification by the SELDI-TOF-MS-based ProteinChipOR System", Nature Methods, 2, 2005.

[14] M. Merchant and S. R. Weinberger, "Recent Advancements in Surface-enhanced Laser Desorption/ionization-time of Flight-mass Spectrometry", Journal of Inter Science, 21(6), 2000.

[15] R. Aebersold and D. R. Goodlett, "Mass Spectrometry in Proteomics", Chemical Reviews, vol. 101, 2001.

[16] R. H. Lilien, H. Farid, and B. R. Donald, "Probabilistic Disease Classification of Expression-Dependent Proteomic Data from Mass Spectrometry of Human Serum", Journal of Computational Biology, 10(6), 2003.

[17] E. F. Petricoin, A. M. Ardekani, B. A. Hitt, P. J. Levine, V. A. Fusaro, S. M. Steinberg, G. B. Mills, C. Simone, D. A. Fishman, E. C. Kohn, and L. A. Liotta, "Use of Proteomic Patterns in Serum to Identify Ovarian Cancer", The Lancet, 16(2), 2002.

[18] G. Ball, S. Mian, F. Holding, R. O. Allibone, J. Lowe, S. Ali, G. Li, S. McCardle, I. O. Ellis, C. Creaser, and R. C. Rees, "An Integrated Approach Utilizing Artificial Neural Networks and SELDI Mass Spectrometry for the Classification of Human Tumours and Rapid Identification of Potential Biomarkers", Journal of Bioinformatics, 18(13), 2002.

[19] S. Weinschenk, P. Jamar, S. C. Yeo, and P. Jamar, GUI Design Essentials, Wiley, 1997.

[20] H. Bensmail and A. Haoudi, "Postgenomics: Proteomics and Bioinformatics in Cancer Research", Journal of Biomedicine and Biotechnology, 2003.

[21] National Cancer Institute Data Bank.

[22] Te-Ming Huang, V. Kecman, and I. Kopriva, Kernel Based Algorithms for Mining Huge Data Sets, Springer-Verlag, 2006.

[23] Z. Chaczko, R. Klempous, J. Nikodem, and J. Rozenblit, "24/7 Software Development in Virtual Student Exchange Groups: Redefining the Work and Study Week", ITHET 7th Annual International Conference, Sydney, Australia, 2006.

5

Worldwide Teams in Software Development

Pawel Cichon*, Zbigniew Huzar†, Zygmunt Mazur‡,
and Adam Mrozowski§

Wroclaw University of Technology, Wroclaw Poland
**pawel.cichon@pwr.wroc.pl,* †*zbigniew.huzar@pwr.wroc.pl,*
‡*zygmunt.mazur@pwr.wroc.pl,* §*adam.mrozowski@pwr.wroc.pl*

Abstract

Worldwide Teams in Software Development describes a unique, adaptive production process and organizational structure of a development team involved in 24/7 software development process with the use of SPEM profile of the UML language. The organization of a company presented in this paper is based on the neutral matrix model including elements essential for remote execution of adaptive projects.

The proposed adaptive process with dedicated organizational structure solves the most important problems that appear in 24/7 development. The presented solution enables organizations to react more quickly to changes during execution, minimize the risk of incorrect functional requirement specifications and to improve the final product's quality through strong verification at every phase of the productive process. The strong support of knowledge transfer between distributed teams makes the described solution the suitable framework for 24/7 development.

Keywords: Teamwork, Worldwide Collaboration, UML, Software Process Engineering Meta-model (SPEM), Open Management Group (OMG)

5.1 Notation

The presentation of research described in this chapter is based on the Software Process Engineering Meta-model (SPEM), which is used to describe a concrete software development process or a family of related software development processes. An object-oriented approach has been chosen to model a family of related software processes and the UML profile has been used as a notation. This introduction covers only a minimal set of process modeling elements necessary to describe these processes. More details can be found in SPEM specification delivered by OMG [11]. At the core of SPEM is an idea that a software development process is a collaboration of abstract active entities called process roles. These process roles perform operations called activities on concrete, tangible entities called work products. The overall goal of the process is to bring a set of work products to a well-defined state. The fragment of SPEM conceptual model is presented in Figure 5.1.

The notation used in the chapter for the presented conceptual model includes the following elements.

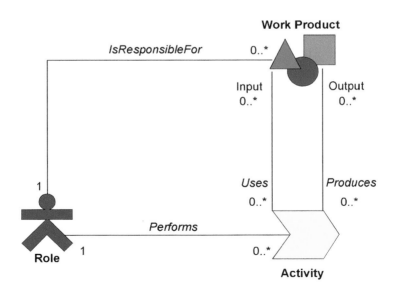

Fig. 5.1 The SPEM conceptual model: Roles, work products, and activities.

Table 5.1 Notation elements.

Symbol	Description
	This symbol indicates a stand alone, complete, end-to-end process class - Process. It is distinguished from normal ProcessComponents[1] in that it is not intended to be combined with other components. The class Process can also represent a family of processes, which allows for multiple overlapping processes to be defined.
	ProcessRole is a subclass of ProcessPerformer described below.
	ProcessPerformer defines a performer for a set of WorkDefinitions in a process. ProcessPerformer has a subclass ProcessRole. ProcessPerformer represents the "whole process" or one of its components in an abstraction, and is used to own WorkDefinitions that do not have a more clearly specified owner.
	Phase is a specialized subclass of WorkDefinition whose prerequisite defines the phase entry criteria and its goal (often called a "milestone") defines the phase exit criteria. Phases are defined with the additional constraint of sequence; that is, their enactments are executed with a series of milestone dates spread over time and often assume minimal (or no) overlap of their activities in time.
	WorkDefinition is a class that describes the work performed in the process. Its main subclass is Activity, but Phase is also a subclass of WorkDefinition.
	Activity is the main subclass of WorkDefinition. It describes a piece of work performed by one ProcessRole: the tasks, operations, and actions that are performed by assisted by a role.
	WorkProduct is anything produced, consumed, or modified by a process. It may be a piece of information, a document, a model, source code, and so on. WorkProduct describes one class of work product produced in a process.
	Document is a category of WorkProduct.
	UMLModel is a category of WorkProduct.

5.2 Introduction

Complex information projects are difficult to estimate because of multi-resource fullness and a short life cycle, which indicates a short period between the next modifications or extensible development of application functionalities [1]. Therefore complex systems are often adapted to budget during the

---

[1]ProcessComponent constitutes a chunk of process descriptions that are internally consistent and may be re-used with other ProcessComponents to assemble a complete process.

process of execution, even though in a canonical model of project management, a business contract is entered into, on the basis of the cost of subsystems and tasks, meaning that the project is decomposed. Most important is to create an information system that makes business run as efficient as possible, and fit within the time frames and the client's budgetary constraints.

Cost pressures have moved today's software development to remote execution. It appeared that remote work may be much cheaper without a negative impact on quality. However, with geographically distributed teams, a second issue — time - could be solved. It is possible to coordinate the work of software development teams in such a way, that every team continues the work of a previous one in order of time zones.

In today's fast changing business environment, the static requirements specification is no longer usable. Changes are common in every IT (Information Technology) project. These changes can easily be handled in a centralized organization structure. There are also several software development methodologies (i.e., SCRUM) in existence that support execution in changing environments. Unfortunately existing solutions cannot be used in changing and distributed environment like the 27/7 development process. In such environments there are additional elements that must be handled. The most important is the knowledge transfer between distributed teams. The knowledge transfer concerns two elements. The first one is the transfer of information about the latest changes and actual requirements status. The second one is the procedure of passing actual work results to the following team.

24/7 development authors suggest using the adaptive development methodology described in this chapter, especially in the case of developing complex systems for business customers. The organizational structure described in this chapter has been already successfully implemented in 24/7 projects executed by SAP. The described adaptive development method used by authors in several commercial 24/7 IT projects turned out to be successful in the case of the "trusted advisor" relationship between the customer and the developer. The reason for this is that the development method assumes that the contract is signed mostly on the base of non-functional requirements with most important elements that expose main priorities: scope,[2] cost and time. This

---

[2] Scope defines specification of business processes that should be covered by the entire project.

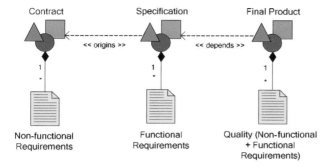

Fig. 5.2 Requirements in the context of adaptive project life cycle.

type of contract is called adaptive contract,[3] and information projects based on this structure are thus called adaptive projects. These projects, as with other information projects types[4] such as waterfall, agile, optional scope etc., are successful, when they meet functional and non-functional requirements, which determine the quality of the final product. The difference is that adaptive projects emphasize extensively non-functional requirements. The non-functional requirements constitute the basis for functional specification, which is postponed to further phases of the adaptive project (Figures 5.2, 5.3 and 5.5).

The authors have chosen such a solution, after trying the RUP (Rational Unified Process) [13] for 24/7 development. In the distributed 24/7 environment the additional workload for transferring changes that appeared during the development process to every distributed team was growing linearly with the complexity of the teams' network. The second reason was that 24/7 development is usually used in projects where time plays the key role. There is no time to create the complex functional requirement specification so the authors decided to create the requirements specification simultaneously with

---

[3]Terms adaptive contracts and adaptive projects used by the authors are also called fixed price contracts and fixed price projects for the sake of stable non-functional requirements (the price set by the client covers resource function, project complexity and its time duration). Adaptive projects are fixed price projects, which allow specifying functional requirements during entire project life cycle (adapting functional to non-functional requirements), this is the reason for "adaptive" adjective usage. Adaptive contracts are fixed price contracts in view of adaptive projects (term adaptive contract is used mainly for syntactic coherence with adaptive project term).

[4]Projects are named mainly for the sake of the most important characteristic, which is the basis for non-formal project typization.

the implementation. This way they are able to take advantage of the shifts in time zone for not only implementation but also for specification.

Specification of functional requirements during project execution poses the main difficulty for software developers because continuous changes of requirements cause difficulty in foreseeing the entire project's realization cycle at the beginning. Therefore a software developer has to respond as quickly as possible to changes at the project execution and functional requirements must meet the client's actual needs with reference to production capabilities, cost and time limits.

The chapter presents a description of the adaptive type of a 24/7 information project, unique execution process of an adaptive project and a company's organizational structure that supports the continuous knowledge transfer and allows remote adaptive project execution with the use of SPEM language [4, 11]. The originality of the described approach relies on the parallel nature of implementation and the design of productive process sub-phases (Figure 5.5). Additionally, it lies in the context of evolving functional requirements and in setting consultants and project data set in a standard matrix organization structure of an information company.

The described solution allows for the maximization of 24/7 advantages. The implementation as well as the specification can be continually extended by distributed teams. The presented solution allows quicker reaction to the functional changes that may appear during the development process.

5.3   24/7 Method as Telecommuting and Virtual Teams Based Project Realization

The rise in the importance of information, the progress in technology enabling the free flow of information and looking for profits were major contributing factors in the appearance of modern corporate organizational structures. These structures allow for flexibility, diffusion and, more importantly, task orientation. The establishment of virtual organizations and telecommuting oriented firms was not a revolution but an evolution. The evolution of organizational forms can be divided into six forms listed below [2]:

1. Centralized enterprises,
2. Partially centralized enterprises,
3. Distributed enterprises,

4. Outsourcing based enterprises,
5. Telecommuting based enterprises,
6. Organizations.

IT companies looking to optimize their share of the software development market searched for solutions allowing them to maximize their profit by expanding their business — that was the main reason for the distributed 24/7 project realization idea, which demands telecommuting.

One of the first telecommuting practices [15] was run by Jack Nilles, the author of The Telecommuting — Transportation Trade off [14], a work acknowledged as the authoritative compendium on this subject. Alvin Toffler is also considered a pioneer in the field of telecommuting and his work The Third Wave [16] was very influential on the subject.

The essence of telecommuting is to perform work duties outside of a central office while communicating and collaborating via teleinformation technologies. Telecommuting requires the formation of virtual teams, which[5] consist of employees connected by a common aim, the project's big picture while separated by distance and thus communicating through the use of teleinformation technology. Telecommuting allows for numerous experts to work in parallel from remote locations and companies to take advantage of efficiencies gained through the use of inexpensive specialists. Telecommuting also has disadvantages, it creates new requirements like the big picture maintenance, which is particularly difficult on large projects, where every worker is saddled with a part of the project and thus the project has a high reliance on the competence of and self reliance of each individual.

In regards to the term virtual team defined above, virtual organizations mean firms that utilize virtual teams. Such a definition of virtual organizations shows their close connection with telecommuting. The work connection factor of virtual organizations is the information flow. The information should be the final product of the cooperation, because only immaterial goods can be transmitted using teleinformation networks [17, 18].

Virtual organizations represent a big opportunity, especially for small and middle sized enterprises, which are able to build this way the modern and worldwide organizations and compete with the biggest on the market. Such a

---

[5]Big picture means complete information data set, which describes the project realization.

solution gives the opportunity to produce advanced systems, which could be unreachable for single middle ranged companies and conventional centralized teams [19]. The main virtual organizations types are [20]:

- **Virtual enterprises:** Differ from conventional (centralized) organizations with much bigger assets distribution. The important difference is various and very wide localization of project teams, which are situated in different parts of the world, but are all concentrated on one production process.[6]
- **Virtual alliances:** Alliances are realized for specific order execution, which used in concert, exceed the possibilities of every single alliance member firm. Usually virtual alliances are built with use of precise formed outsourcing contracts.

5.4 24/7 Development Process

5.4.1 Success Prerequisites in 24/7 Development

24/7 projects completed by authors to date, were related to ERP [21] systems development for large customers.

During the first project, teams had to cope with many organizational problems, caused by the 24/7 idea. The problems were related to ever changing customer requirements, cooperation in remote environment and knowledge transfer between teams.

The today's business world is changing quickly. Prerequisites and assumptions taken on one day may be not valid the next. Traditional development methodologies like RUP or PRINCE [22] are often useful for large projects. Unfortunately those are long lasting activities that cannot be used under time pressure in projects where changes in requirements are common. There are solutions supporting such a changing environment and fast development like SCRUM. However, some of its activities, like daily project meetings of a whole team make this solution unacceptable for many remote teams as well.

Authors also noticed that human elements and cultural differences have a big impact on work efficiency especially in 74/7 development. People who

---

[6]Production process means tasks sequence.

work together but don't know and don't see each other easily blame the others for all errors.

In the changing environment another key element came into play. There appeared problems with changed requirements specification. There are many solutions, i.e., CVS (Concurrent Versions System) [23], supporting remote cooperation on one source code. The first 24/7 project showed that code can be easy distributed but specifications cannot. Even if it is in the form of a daily updated document the distributed developers often miss updating or checking it. So authors did not experience any problems with transferring knowledge about the source code or technology but there appeared serious problems with transferring knowledge about actual requirements.

The method and organizational structure presented in the chapter represent answers to these problems. Authors developed and put into practice 24/7 adaptive IT systems development method and created a unique organizational structure for their teams that solved many of the aforementioned problems. To make 24/7 development successful the assumptions described below should be considered.

5.4.1.1 Right people in distributed teams

Getting the right people to the team is one of the most important tasks in the whole process of 24/7 software development. The organizational structure of distributed teams designed for 24/7 development is described in details in Chapter 5, but every structure must be filled with people. Pat Riley the famous NBA Coach used to say that the best coach was the one with the best players. This sentence fits well to software development. The best project manager in 24/7 development is always the one with the best people on their team.

Many managers assume that projects can be executed with average people. This is true in the case of standard methodologies like RUP. However, in 24/7 projects, team members must be outstanding. It is the environment that requires very good technical specialists with high level of self-discipline.

Getting the right people does not mean getting the best and the most expensive. The manager should choose and judge the team members. Overqualified members are always useful, but costs of employing them must also be considered. Bary Bohem in [1] suggested recruiting to the team less, but more qualified people. His research indicates that 20% of the best qualified team members generate 50% of the product. So in high risk projects, like 24/7 development, recruiting overqualified member seems justified.

The authors took a part in a 24/7 project, where a dozen of most experienced programmers were chosen. It quickly appeared that technical specialists are not able to cooperate. The solution was to release half of the team to other projects. The rest were supposed to chose partners with whom they wanted to cooperate. Surprisingly the selected persons were not the best specialist but they were very popular in the company. After such a reconstruction the work progress was much better than before and all conflicts disappeared. The above example shows that the best team member is not always the most experienced one. Every member may be valuable in some domains, not only technical, important for the 24/7 project. The 24/7 project manager's task is to chose such people. It's much more difficult and important to select the right boat crew than to decide where to sail. As with 24/7 development, the most important is selecting the team members, not a methodology or organizational structure. The development process may be helpful but even the best process will not allow one to succeed if the wrong people are on the team. Unfortunately many managers don't understand that the process cannot substitute for skills.

All of the above information is also valid for the customer's recruitment of a team. If tone cannot build the right team for the customer, it's better to skip such a project.

5.4.1.2 Understanding the product vision and its business goals

In 24/7 development a manager must ensure that all team members understand the project vision and their role in the project. It's essential for distributed 24/7 development, where sometimes members do not have continuous access to consultants or managers and must take own assumptions. The project vision may contain the architectural system overview and process description. Usually it also contains the component description and interface definitions. Authors assume that the vision may be changed during the project execution but the most important thing is to keep all team members up to date with its actual vision. All members in distributed teams must know the schedule and understand where the critical dates and deadlines come from. The manager or leader in each team must clarify, why some tasks are more import ant and some have a lower priority. The practice showed that the better the team knows the above elements, the more efficient it works and members are able to make better decisions.

5.4.1.3 Quick reaction to changes

Every project contains known and unknown elements. That's why the manager and teams must strike a balance between planning and flexibility. Authors have managed several projects where at the beginning, only the product vision was known. For such a project with high exploration level, there is no possibility for long term planning. This is the perfect environment for 24/7 development. At the beginning of the project only the vision is known, but during the execution new facts appear, and can be incorporated 24 hours a day by the distributed teams. This way, especially at the beginning of the project, the progress in 24/7 development is much higher than in traditional approach.

Authors assumed that changes in a 24/7 project may be common and may apply to every aspect of the product. This is an answer to an actual business situation where changes in corporations are common, if they only want to survive in competitive markets. Corporations must adjust to changing market requirements together with their information systems. In such a situation, the traditional approach of following a preliminary prepared plan is not sufficient. After several months of traditional software development the product may be useless for the customer because of a changed business environment.

5.4.1.4 Frequent prototypes

There are many reports available indicating that large projects for which the documentation is developed for months have much lower success rate than the other projects. Why is it happening like this? The teams that have the complete documentation for the whole project act in a linear way. They follow the plans and the static documentation while everything around is changing. There is no room for introducing interesting new ideas that appeared during the project execution. The described 24/7 development process puts emphasis on delivering the working solution in early project phase, and not the detailed specification. Taking advantage of fast 24/7 product delivery the customer can take several advantages of having early a working demo. The most important is that while having the early working demo his vision of the product may be clarified or changed. The practice showed that the vision is very often changed after the clients see the working product. The changes can only be applied if product is developed in short iterations. Thanks to frequent prototyping the solution delivered in next iterations can be best fitted to actual customer requirements. The reaction to the changes here can be quicker than in standard development projects because of the time zones differences. This way, in many cases, the customer may already test the changes that in the morning that he submitted on the previous day in the afternoon. This seems to be one of the biggest advantages of 24/7 development and so far it has played a key role in convincing customers to leverage the 24/7 idea. Of course documentation cannot be abandoned all together. It's required, and in a described environment technical documentation is produced continuously with the source code in the form of daily closing reports. However the functional requirements are delivered by the consultants gathering the requirements from the customer.

5.4.1.5　Strong interactions between teams

In the mid 1990's, a business process reengineering revolution took place. The result in the beginning of XXI century was that in many organizations' processes became more important than people. In the case of 24/7 development processes should support the team activities rather than to tell them what to do.

With the idea of 24/7 development the world's best developers can be used in one project. Thanks to the team's distribution, they don't need to work in one office. They may even practice at a home office which strongly increases morale. The knowledge transfer in such a structure is done by daily meeting of the whole development team. The programmers working from home must participate in the meetings through videoconferences. Every day each team member can also plan an interview with the consultant working on client's site or with the developers from the preceding time zone. Thanks to these meetings and interactions the distributed teams can easily undertake critical decisions. While taking a critical decision each team member can rely on the knowledge and experience of every member that he/she interacts with.

5.4.1.6　Delivering customer value by every distributed team

The authors encountered lots of companies, for which assurance of accordance with initial project requirements was paramount. This came from the organizational climate of those companies, in which project managers demand that every worker is responsible for reporting or other supportive tasks like accounting etc. In many situations this was the reason that employees paid less attention to do their real (technical) tasks than to reports, which are verified by the managers in an ongoing basis. In the 24/7 model it is very important to put a strict border between supportive and technical tasks, as only this way can project managers put more attention on product development execution technical tasks and properly allocate resources to suitable tasks regarding team members competences. 24/7 projects execution should emphasize the rule — minimum documentation and easy mechanism of knowledge aggregation means fluent and successful execution. This concept of making production process as light as possible came from the term "Lean Manufacturing", which comes from Japan and its motor industry (specifically was used by Toyota Motor Corporation), but was popularized by Americans — James P. Womack, Daniel T. Jones and Daniel Roos [24].

According to Womack and Jones, light production in general 'gives possibility to produce more while using less — less people, less units, less infrastructure, time and places — but together with simulates realization which guaranties delivering product pieces, the client is expecting continuously' [25], and this is possible only by supportive tasks reduction. The authors asked 36 PMI [26] project managers engaged in various software developments projects to estimate, what percent of technical tasks are generally involved in the production process. The answer was — 25%. A very similar analysis made by Allen Ward, but regarding companies using "Lean Manufacturing" gave results — 80%, therefore authors made the decision to use "Lean Manufacturing" principals in software development and look for redundant and supportive tasks which could be omitted in well know methodologies like RUP or USDP (Unified Software Development Process) [27]. Using a proportion of supportive to technical tasks in relation to proper balances verification — with technical task at the highest values, and supportive the lowest — the authors decided to omit the 50% of supportive tasks with the lowest priorities. These tasks were chosen by the clients and tested utilizing the "light" production methodology in a real software development process. As expected, from the team member's point of view, there were fewer duties to do, from the project manager's perspective more administrative work, but less verification of reports and documentation, but most importantly, from the client's point of view, nothing changed. So simply the client didn't notice the lack of some supportive tasks (e.g., lack of reports, documentation, meeting etc.) and from the production process point of view nothing critical happened. Nevertheless such a production process was the reason for giving to the individual team members more responsibility and the right to make decisions regarding task realization. This led to senior team members showing junior members how the task should be worked out and started to do supportive task, like teaching, caring about timelines etc, which in general should be omitted — everyone should be focused on individual tasks, and take care of himself. The conclusion was that "light" processes in the teams of 24/7 production process realization are best to use with experienced staff or with weekly meetings ("brain storms" — video/teleconferences) in balanced teams including juniors and seniors developers. Only this way can junior team members feel comfortable and safe with decisions regarding implementation, which were made together with senior members during meetings.

5.4.1.7 Functionalities prioritization

In executing continuous software development process, the authors used an iterative model of software development. This was extended by the implementation of progressive functionalities, based on priorities. This means that at the very beginning, teams are focused on the realization of basic functionalities (most important from the clients point of view – the client chooses the most important from his point of view, and marks it as priorities), which in the next iterations are evolving. Every iteration should come with expected progress and functionalities realization and is preceded by short support, summary and planning meetings (video/teleconferences). Such meetings motivate teams' members to quickly execute tasks and assure project managers that team members understand allocated tasks. In the 24/7 method, iterations should be as short as possible, because only this way teams have big picture of and control over progress. Additionally meetings which include the planning of the next round of functionalities are very important in terms of distributed project realization and are the most important difference between the traditional iteration model, in which functionalities are established at the very begin of the project and can't evolve. It is very good if a client can take a part in such meetings, only this way teams can be sure, that the extension of functional requirements is required and accepted by the client.

5.4.1.8   Progressive risk reduction

Software development is probably rated the highest possible risk in the whole project management domain. Tom DeMarco and Tim Lister in [28] wrote "if the project does not have a risk — do not undertake it". The risk is related with a wide range of issues. It starts from the technical risk, goes through the marketing risks and finishes on financial risks. Many software development methodologies forget about risk management. Sometimes it is an occasional activity. In 24/7 software development risk management must be closely related with the development process. On every stage of the project the consultant working at the customer's site discusses all functionalities and estimates its business value and the risk associated with undertaking such an implementation. Usually two questions are asked during the discussions:

- Which functionalities have the biggest business value?
- Which functionalities are most risky from an implementation perspective?

These two questions are related with the same problem of generating added value, seen from different perspectives. For example while developing the new software there is a technology risk of using new, unknown technology, but there is also another risk related to implementing the wrong functionalities. Both of these risks must be discussed together by the consultant and the customer.

The iterations in 24/7 development should be planned in such a way that it covers the most risky elements in early iterations. In some iteration the functions using new and unknown technology will be implemented but in the others consultants may opt for implementing a functionality that is easy from a technical perspective, but the customer is not sure about his requirements and needs a working demo to clarify his/her vision. In this case, it is not the technical risk that is high but the risk of changes to the specification. So the risk in 24/7 development is also evaluated based on its value. Sometimes the more valuable is the early technical risk elimination and sometimes it is the functional one.

5.4.2   The 24/7 Development Process

The traditional approach to an information project execution presupposes that the phases of design, implementation, testing and migration occur sequentially.

The next phase starts upon completion of the preceding phase, therefore the first versions of the working system are delivered to the client relatively late because in the last phases of the project life cycle. The adaptive process used by authors for 24/7 development is guided by a different approach to the traditional one, mainly because it delivers the working application to the client as soon as possible. This is achieved through making the design sub-phase parallel to the implementation sub-phase, making them iterative and taking advantage of phase interdependency (Figures 5.2, 5.4). This leads to the creation of a sub-process oriented to the delivery of the project development results to the client in the form of functional subsystems. This way each member from every distributed team can see that the project is moving forward and other teams are also working. There is also a greater value of frequent deliverables. It is knowledge transfer. Working functionality shows programmers the direction that should be taken. Programmers may not know the latest changes in the documentation, but every time new working functionality is delivered they know the business goal of the entire system better. So they are able to undertake more decisions on their own without project manager's confirmation or consultants working at customer's site.

The adaptive method of 24/7 used by authors for complex IT system development allows for functional requirement extension during the project execution. It was proved that a realization of just 20 percent of the functionality makes 80 percent of client's satisfaction [12]; this has implications in the way of quality and business value of the final product. An adaptive project promotes this point of view first of all by emphasizing quick delivery of the most important 20 percent of functionality and the project guidance by the customer. The client verifies and specifies functionality of the system and in such a way affects the whole informational system execution process. Continuous process adaptation to customer requirements in order to obtain the highest possible business value of the final product describes the adaptive process, which is fundamental for adaptive projects.

The analysis of project management methodologies (RUP, USDP) leads to conclude that most of them have no strict marginal conditions.[7] Such an

---

[7]Environment parameters that must be met for safe application of the methodology.

instance brings about solutions that are oriented to predictability[8] in unpredictable processes which increases the risk of project failure. The developers distributed around the world don't see working prototypes too often, so their morale decreases dramatically after several weeks. The adaptive 24/7 projects are partially unpredictable, because of defining functional requirements during process realization and that couldn't signify unpredictable and uncontrolled process (called realization chaos). Therefore a repetitive realization process and appropriate organizational structure has to be created which will assure control over unpredictability, let members react efficiently to changes of functional requirements and allow for easy knowledge transfer while realization of the project. Reaction to functional requirements changes needs tight collaboration between the team members and domain experts (consultants) [10]. This collaboration goes beyond standard business roles (every worker should be regarded as an expert and consultations should constitute a preferred solution of problems) and preparation of an evolving process to meet continuous functional requirement realization needs regular audits (evaluation of milestone progress).

The 24/7 project execution process starts with the vision phase. In that stage consultants and customer representatives focus on, among other things, defining a document describing non-functional requirements (Figure 5.3).

This activity must take place at the customer site where business consultants cooperate with the customer team while creating non functional requirements and specifications. The consultants, working together with domain experts, create a list of business processes that have to be covered by a system. It is important to describe points of integration of the new system, existing infrastructure and external processes. The goal of the vision phase is to define the project scope in the context of non-functional requirements and point out the most important threats. Furthermore, there is a coordinator and domain experts (on the customer's side) selected, responsible for championing the development of functional requirements for the system. They decide about supplementing or removing specified functionality proposed by the consultants. These non functional requirements are the entry point for remote cooperation. From this point the distributed teams can start their work and the advantage

---

[8]Predictable solutions are the solutions that have all functional and non-functional requirements set before the project starts. This allows for creating of an entire project at the very begin.

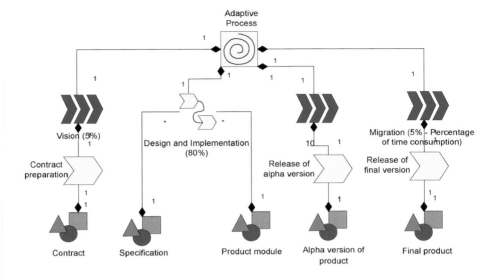

Fig. 5.3 The execution process of adaptive projects with marked phases and artifacts.

of time zones shift can be realized. The non functional specification goal is to show the general business goal of the system to every programmer involved in the development process. Knowing the system goal and its business value for the customer, they may focus on the most important activities and undertake better decisions while working remotely without continuous access to the customer.

The method presented here assumes an iterative approach regarding the risk[9] in the design and implementation phases. An extensive and narrow[10] implementation[11] of many components[12] [7] come true in parallel. This way the advantages of time zone shifts can be taken. The objective is a quick stabilization which entails partial implementation of base software and hardware components, associated with functionality that has high business priorities and definition of interfaces the cooperation rules and environment. From the customer's viewpoint the approach above has a psychological value because

---

[9]For the early iterations the most risky elements are chosen.

[10]Implementation that contains all the most important elements of the system from the customer and supplier's perspective.

[11]Programming activities that include only frames of functionalities. These frames may be extended in further iterations.

[12]Subsystem that contains one or several coherent use cases.

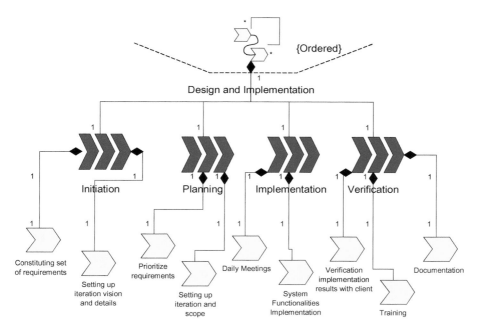

Fig. 5.4 Decomposition of a design and implementation phase into sub phases and activities.

early implementation of key functionalities helps to stabilize the high level of customer's involvement in the project and that facilitates smooth project execution [3]. It is very important because in iterative designing directed to the client, the customer decides about elements introduced in successive iterations, demanding implementation of the most important modules from his/her point of view. This is also helpful from the remote team's perspective, because developers can see working elements of the system and can clarify their vision of a whole product. So the world distributed development teams take advantage of using the described method as well as the customer. A single iteration[13] consists of sub phases, presented on Figure 5.4.

The most important activities in specified sub phases are as follows:

5.4.2.1   Initiation

The team members create a list of use cases and their extensions. Moreover, all the possible risks should be identified. The sub-phases' goal is to specify

---

[13] Set of activities that lead to creation of a functional system module.

the functional elements of a system that should be implemented in a coming iteration.

5.4.2.2 Planning

The choice of functionalities (their verification) that must be implemented in the context of business value for a client. Setting the iteration scope by removing less important functional requirements from the planning phase. Also there is no planning based on Work Breakdown Structures (WBS). The authors used Features Breakdown Structures (FBS) in their projects. The standard WBS had been adjusted to fulfill the 24/7 requirements and tasks in WBS had been replaced by functionalities in FBS. This allowed the team members to focus on working functionalities. Authors noticed that it's much easier to manage functionalities than tasks in distributed development. Moreover if from any reasons the task was not completed by the developer in an earlier time zone it's difficult to introduce the task to someone one else in the following time zone. It's much easier to introduce new functionality.

5.4.2.3 Implementation

Putting into practice those functionalities which are consistent with the verified requirements list. The sub-phase ends with a release of the internal[14] tested version of a subsystem.

The FBS should be detailed enough to cover functionalities that can be implemented by one programmer within a few days. If this is not possible and several programmers must work on one functionality; the essential is to end a day completing and testing at least one working element of the functionality before passing the work to the programmer in the next time zone. This assumption leads to the realization that a remote programmer's day is not the same length every day. In the projects completed so far within the 24/7 framework, developers had to work varying amounts of time, ranging from 10 hours a day in some cases and sometimes only 6, depending on the functionality they implemented. Of course if there is a functionality that is not related with any other it can be developed in one's spare time.

---

[14] Internal testing means verification of partially finished product and is made during production process (Verification sub-phase of design and implementation phase).

5.4.2.4 Verification

Representatives of the developer and the client organize a meeting to discuss the implemented functionality. They point out weaknesses and strengths of a solution, identify potential problems and indicate possible extensions. The goal is to show the best possible way to develop the existing application. This activity must be performed at a customer's site and cannot be done remotely.

The 24/7 project should always start with an event where all developers can meet themselves. The practice showed that this has a huge impact on the future performance. People work much more efficient if they personally know with whom they cooperate. This is the only chance to integrate the team so the value of this activity is very high. For longer projects such as events, this could be planned every six months for example.

The whole project is executed with the use of iterations (Figure 5.5). In 24/7 development it is essential to do the two first iterations in a centralized structure at one site (Figure 5.6). A leaders' team should be created for these iterations. A common team should contain one or two of the most experienced

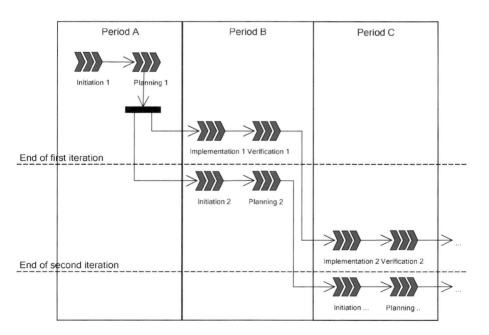

Fig. 5.5 Parallelism of iterations of design and implementation sub-phases.

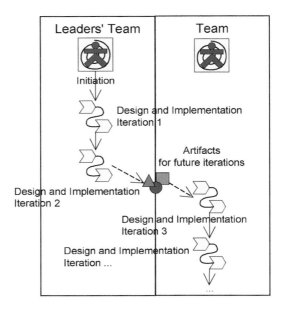

Fig. 5.6 24/7 iterative implementation based on distributed teams.

programmers from the future remote teams. During these two iterations the greatest emphasis is put on requirement specification, which includes the most important components and their interfaces. The components related to the most important for the customer's business are created more efficiently due to direct communication and cooperation of the best programmers. When this basis is ready, the main programmers are coming back to their teams (here is where the 24/7 development starts) and acting as team leaders. This way every team has its leader but also a person who has created the fundamental elements of a system. Moreover, thanks to the team leaders' direct cooperation in the first two iterations further communication between teams has also improved during the whole project. Leaders know each other and by common work at first two iterations some cultural barriers can be removed.

At the end of a working day each developer is responsible for document-ing and testing his work. This shouldn't be a detailed UML specification of completed work. It should be a short task, not longer than 20 minutes, during which a report describing the completed functionality should be created. If the functionality was not completed by the previous team only the completed functions should be described. The closing report must contain the following elements:

- Function name and location.
- Input and output parameters.
- Function description.
- Usage example.

From the commencement of the third iteration, the working day of the remote team starts with becoming familiar with the closing reports from the previous team. After 1 hour a team meeting is organized. Until this time, all members should already know what are they supposed to do during the day and what was previously done by others. The meeting shouldn't be longer than 30 minutes and all team members must take part. Each member must loudly articulate what he did on the previous day, what was done by other remote teams and what he/she plans for the actual day. This practice used by authors in 24/7 development helped to identify possible problems and misunderstandings. If any developer had problems with articulating his tasks the team leader was responsible for explaining him the work to be done. This method promotes the approach that wrong decisions are better than no decisions. Wrong decision can be corrected in the future iterations but no decision leads to the project stagnation [8]. The tasks set for one day must be completed during that day. If any problems appear and it's impossible to complete even a single function on one day it must be explained in the closing report and all not working source codes must be removed. This way every following developer will have a working application and closing report containing source codes and problems identified by the previous developer.

Adding a new functionality to an ongoing iteration has to be locked. Consultants, team members or the customer's representatives are not allowed to add new functionalities to the running iteration. Team leaders together with the project managers are acting as a firewall that isolates the team members from customer's change requests. If changes appear, the project managers negotiates their implementation in the future iterations (putting into tasks queue) [9]. 24/7 development is a specific environment in which developers should be isolated from any political problems and customer.

The last iteration in the design and implementation phase must end with the release of the final, tested alpha version of the system. Once this is successful, this project moves to the validation phase. The customer can now test the whole system and report errors. Requests for supplementation or removal of any functionality cannot be reported on this level. If a necessary change requests occurs, they may be written as an extra task. After the external[15] testing has been completed and the errors corrected, the system is transferred from a test environment to production servers. This stage is called migration. The advantage of 24/7 development here is that the productive application start can be 24 hours supported by distributed teams. No team relocation is needed if the customer is in a different time zone. Such a productive starts usually take place at night when the servers are not used. During this productive starts the teams should be available at all times. It was a very stressful task before the 24/7 development was introduced. Some developers had to spend nights at work supporting the productive start. With the 24/7 development, one team in the world is always available in its normal local working hours.

An internal version release at the end of each iteration and a demonstration version for the client at least every two iterations should be created to clarify the product vision for the customer as well as the developer.

5.5 Software Developer Organizational Structure for 24/7 Projects Execution

The organizational structure presents a network of organization bonds[16] and dependencies, which constitute a connection among all company resources, both human as well as material. Moreover, it describes distribution of work activities and employees' hierarchy, shows correlation between functions and activities and specifies a responsibility allocation. The following bonds occur in organization [2]:

- **Technical (task oriented):** These bonds connect positions that are involved in task execution (implementation). Following stereotypes shown in Figure 5.7 ≪produces≫, ≪elaborates≫.
- **Functional (function oriented):** These bonds connect positions that are involved in decision management and tasks planning.

---

[15]External testing is synonym to validation and means tests performed by customer after final product release by software developer.

[16]Information exchange channel.

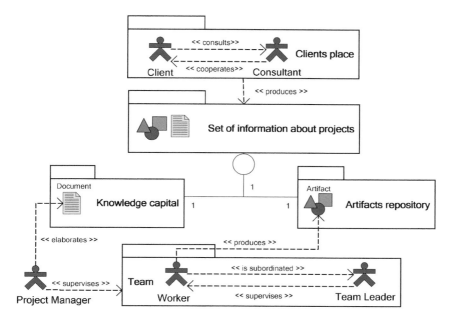

Fig. 5.7 A neutral matrix model for presented process of 24/7 projects execution.

Following stereotypes shown in Figure 5.7 ≪supervises≫, ≪is subordinated≫, ≪consults≫, ≪cooperates≫.

The method of adaptive 24/7 project execution presented in the previous chapter realization is efficient only when it is based on an appropriate organizational model. Organizational models that work best for the presented process are task-oriented structures like the matrix model [2]. The matrix model, depends on managers' and team leaders' competences, can be divided to: weak — when a manager allows the team leader to allocate tasks to the team members, strong — when the manager assigns the tasks and neutral — when the manager together with team members perform task allocation. Whatever the model, the manager's role should always be focused on planning, scheduling and budgeting and the team leaders' role on technical supervision. Figure 5.7 presents an organizational model drafted for remote run of process presented in the previous chapters and is based on neutral matrix model.

A team member is directly subordinate to the team leader, who verifies the work progress and is responsible for implementation quality. Moreover, leaders are responsible for an implementation's coherency, repository data update and team members' training. As presented on picture 7 there should be a consultant or an architect working all the time at the customer's site. During the commercial 24/7 project execution it occurred that one consultant can handle only two remote teams. In the case of more distributed teams there might be more consultants working with customers as necessary. The consultants together with the customer create requirements specification and are responsible for transferring it from customer to the teams. The knowledge transfer is not the transfer of source codes and work progress. This is handled by the daily closing reports and CVS systems. The knowledge transfer done by on site consultants is the transfer of user requirements. So the structure may be compared to a 'client-server' functionality where the onsite consultant is a server for a remotely working development teams. Every request from the development team must be sent to the consultant who is responsible for handling it and generating response.

Project managers are responsible for scheduling, budget verification, changes and risk management. They may also allocate tasks packages to teams with the cooperation of team leaders. The executed assignment is based on functional requirements and system decomposition delivered by the consultants working at a customers' site, who are responsible for the system analysis and decomposition into modules.[17] Project managers, consultants and team leaders are obliged to report work progress at weekly project meetings. The weekly project meeting can of course be done remotely. The meeting looks the same here as the daily development meeting. Each consultant and leader must articulate what was done in the previous week and what are the plans for the next. This helps to create a common product vision for both consultants and team leaders.

The set of information about projects presented on picture 7 includes knowledge capital and remote repository, which is extremely important for remote cooperation amongst teams and consultants. Firstly, it allows teams to keep conscious of the project's big picture; secondly, it's essential for future projects and maintenance of already-designed systems. Thirdly, it's an integral part of intellectual capital of an organization. It contains the knowledge about previous projects that could be used for result comparison,

---

[17] A functional part of the system.

drawing conclusions and risk mitigation. The knowledge capital covers project documentation — schedule, budget, cost tables, risk calculations, teams and resources lists, daily closing reports and other data important from the managing perspective. The remote repository is a set of artifacts — product source codes with descriptions, managed by a versioning system with change annotations. Every worker must have uninterrupted access to the set of information about the projects, as presented in Figure 5.7. This is the only way in which the workers may understand their role in the project while working remotely and localize assigned tasks in the perspective of the project as a whole.

5.6 Summary

The developer as well as the customer takes on the advantages and risks of 24/7 development projects. There aren't any management methods dedicated to these kinds of development projects and the practice shows that the application of universal solutions to 24/7 production processes leads to project failure. Due to the development process proposed in the paper being adjusted to distributed teams, its orientation to changes of functional requirements the managers may control the development process more efficiently and the software developer may react more quickly to the customer's requirement changes. The presented solution proposes a modern organizational structure for every project execution, which ensures stable and clear communication between the remote teams and the customer, essential for quick response to changes and to ensure developmental consistency. Thanks to this and directly to consultants working at the customers' site, the team members have continuous access to the functional requirements and domain experts, even when working remotely. The knowledge transfer between teams and developers is done using daily closing reports, which are at most, a one page long document summarizing work done by every developer at the end of his workday. The second layer of knowledge transfer is the requirements transfer from consultants to development teams. This solution ensures that every developer in the world will always have actual requirements specification and actual source code. A focus on functionality as opposed to tasks (WBS is replaced by FBS) helps developers undertake their own decisions based on the goal of the system.

The described method has been successfully used in several commercial projects where complex ERP systems had been continuously developed by programmers from China, USA, Germany and Poland.

References

[1] B. Boehm, Software Engineering Economics, Prentice-Hall, 1981.

[2] P. Cichon and Z. Huzar, Remote project management using modern organizational forms, ISAT Proceedings, Wroclaw University of Technology Publishing House, pp. 39–47, 2004.

[3] A. Cockburn, Characterizing People as Non-Linear, First-Order Components in Software Development, http://alistair.cockburn.us/crystal/articles/cpanfocisd/characterizing peopleasnonlinear.html, 2003.

[4] I. Dubielewicz and J. Sas, SPEM/UML Profile in business processes of project management, WNT, pp. 479–495, 2003.

[5] J. Highsmith, Agile Project Management: Principles and Tools, Cutter Consortium Executive Report, 2002.

[6] J. Jonson, Standish Group Chaos Report, 2001.

[7] A. Kasprzyk, SelectPerspective™ Methodology, Components in practice, WNT, pp. 121–126, 2003.

[8] A. MacCormack, Product-Development Practices That Work, MIT Sloan Management Review, 2001.

[9] A. MacCormack, R. Verganti, and M. Iansiti, Developing Products on Internet Time: The Anatomy of a Flexible Development Process, Management Science, 2001.

[10] S. McConnell, Software Project Survival Guide, Microsoft Press, 1998.

[11] OMG, Software Process Engineering Metamodel Specification, OMG Press, 2005.

[12] M. Poppendieck, Website: www.poppendieck.com, 2004.

[13] Rational Unified Process, IBM Corporation Website: http://www.ibm.com/software/awdtools/rup/.

[14] J. Nilles, The Telecommuting — Transportation Trade off, 1990, 1992–1993.

[15] European Telework Development (ETD), http://www.eto.org.uk/etd/, 2002.

[16] A. Toffler, The Third Wave, 1989

[17] Sky et al., Tele-information Networks and Communications 1999.

[18] Hay et al., Collaborative Enterprise Environments, 2002.

[19] Mol et al., Mid-Ranged Companies and Conventional Centralized Teams.

[20] K. Wac, Virtual Enterprises, Virtual Organizations, 2002.

[21] Enterprise resource planning (ERP), http://www.erp.com/.

[22] Prince 2: Project Management Methodology and Standardization, http://www.apmgroup.co.uk/PRINCE2/PRINCE2Home.asp.

[23] Open Source Version Control System (CVS), www.nongnu.org/cvs/.

[24] D. Roos, J. Womack, and D. Jones, Lean Manufacturing Process, 2002.

[25] J. Womack and D. Jones, Lean Thinking: Revised and Updated, Simon & Schuster, New York, NY, 1999.

[26] Project Management Institute (PMI) Process Engineering: http://www.pmi.org/.

[27] Unified Software Development Process (USDP), IBM Corporation, http://www.ibm.com/developerworks/rational/library/5359.html.

[28] T. DeMarco and T. Lister, Project Management Domain Peopleware, 1976.

6

Virtual Student Exchange: Developing New Educational Paradigms to Support 24-7 Engineering

Eckehard Doerry* and Wolf-Dieter Otte[†]

Department of Computer Science, Northern Arizona University, Flagstaff, Arizona, United States of America
*eck.doerry@nau.edu, [†]dieter.otte@nau.edu

Abstract

The increasing globalization of corporate economies has changed the face of engineering practice. In addition to core engineering skills, modern engineers working under the "24-7 engineering" paradigm must possess cross-cultural communication skills, team management skills, and the ability to perform effectively within geographically distributed teams. In this chapter, we review our experiences in exploring a novel paradigm for teaching this skill set within a novel internationalized curricular model that forces students to engage the challenges of 24-7 engineering in a direct, hands-on fashion. We begin by describing a novel curricular paradigm we have been exploring called the Global Engineering College (GEC), that is based on the idea of seamlessly combining the curricula and educational opportunities of several internationally distributed engineering institutions to create a virtual engineering college spanning multiple countries and cultures. In the second half of the chapter, we review our experiences piloting a key element of the GEC model most relevant to the theme of this volume, namely the concept of Virtual Student Exchange (VSX), which centers around medium-term collaboration of students on joint projects within virtual development teams distributed across several international institutions.

Keywords: Virtual Student Exchange, Global Engineering College, Educational Paradigms, 24-7 Engineering, Curriculum Internationalization.

6.1 Introduction

For the past several decades, the internationalization of college curricula has been a prominent theme in discussions of curricular reform in higher education [1, 2, 3, 4, 5, 6, 7]. Few question the necessity of this reform, and the rapid progress of globalization during the last ten years has lent new urgency to this need [8, 9]. A number of institutions have taken concrete steps toward implementing internationalization within individual academic units as well as across the university as a whole. As early as 1993, Oregon State University introduced its "Passport" International Degree Program under which students can supplement a degree in virtually any field with an "international degree [10]". At about the same time, the University of Rhode Island began offering a Bachelor's Degree in International Engineering, a five-year program that graduates students with a traditional engineering degree as well as a B.A. in a language [11].

While the initial impetus for internationalization may have come from humanities and the social sciences (the traditional study abroad disciplines), engineering and the natural sciences have realized that their graduates also require strong international skills in order to succeed in the global engineering workplace of the twenty-first century.

In April 1995, the cover story of PRISM, the journal of the American Society for Engineering Education (ASEE), referred to over 70 engineering programs with international components [12]. Since then, the rationale for such programs has only grown stronger; the world's economy has become vastly more interdependent, exports account for an increasing percentage of economic activity, and capital, work and jobs move rapidly and frequently from one continent to another. As a result, the trend towards smaller, more independent collaborative development teams over the last two decades of modern engineering practice has rapidly evolved into international collaborative teaming. Recent cover stories in ASEE's PRISM explore the effect of these trends on modern engineering practice [13, 14, 15]; the overall conclusion is unanimous: all recent engineering graduates can expect to work, at some point their careers, on teams with members from varied cultural and linguistic backgrounds; increasingly, these teams are geographically distributed as well, spanning national, cultural, and linguistic boundaries.

This new global model of engineering practice, with teams distributed around the world all working together on a common work product, lies at the heart of the 24-7 engineering paradigm. As emphasized by other chapters in this collection, the challenge of initiating, coordinating, and managing such a distributed development effort is substantially greater than with mundane co-located projects. In addition to the usual set of project management skills, new skills are required to deal with differing time zones, synchronize differing cultures and work styles, effectively organize tasks and completion status across two or more independent working groups, and to fluidly manage passing project development off to distant colleagues on a daily basis.

The central motivation for the initiative described in this chapter is that (a) it is critical to train future engineering graduates to work effectively in this global, distributed design and development context, and (b) that, like traditional project management skills, the only true way to learn these extended skills is by practicing them in a realistic educational context. As we began working with our international partners to develop a curricular model for such distributed teaming experiences, however, we soon realized that what was needed was not just one small change to add an international teaming experience in a single course, but a comprehensive rethinking of the current engineering education paradigm in the United States to integrate global perspectives and training experiences into every aspect of modern engineering education.

6.2   The Global Engineering College

To make international engineering training more relevant and accessible to all engineering undergraduates, we have developed a novel curricular model for engineering education called the Global Engineering College (GEC) that injects international perspectives into every aspect of the curriculum. In addition to comprehensive internationalization of our engineering curriculum, the

GEC concept aims to leverage recent technological developments to essentially create a single "virtual" engineering college that integrates selected NAU courses with parallel courses at our partner institutions abroad. Students at one university will be able to participate via Internet in design courses offered at any partner university. An important side benefit in this age of dwindling educational resources is that students will have access to the full array of specialized elective topics, laboratory equipment and practical experiences available at any partner university.

From a practical perspective, the GEC model consists of four key elements that interact in complementary fashion to provide a wide range of international experience and training opportunities:

- **Curriculum Internationalization:** International perspectives can be integrated into existing engineering course curricula by replacing generic, context-free assignments and projects with "scenario-based" challenges, in which the same pedagogic exercises are situated in international contexts. For example, rather than being asked to "design a bridge to such and such specifications", students are asked to design a bridge in a specific foreign locale, taking into consideration international issues like materials, measurement differences, currencies, local availability of capital and labor, while still exercising conventional core engineering skills.
- **Virtual Student Exchange (VSX):** Students at NAU and its partner institutions abroad participate in each other's design courses at a distance by leveraging advances in a number of key internet technologies. This allows us to "bring the world into our classrooms", ensuring that even students who never go abroad are exposed to international teaming and collaboration; NAU students gain access to a wide range of curricula offered at participating institutions. VSX is also the core mechanism for training students to operate in the distributed design and development scenarios that lie at the heart of the 24-7 engineering paradigm.
- **Global Internships:** Prerequisites for success in modern international corporate environments include sensitivity and adaptability to differences in work habits, differing legal environments, and respect for local customs and mores. Because these types

of experiential knowledge can be best gained in an international workplace, streamlined access to a global internship experience for motivated students is an essential part of the Global Engineering College model.

- **Engineering-specific Language Instruction:** Because generic university language courses do not take into account the specialized needs of engineering students, we have developed a model for accelerated, engineering-specific language instruction that will, within a single year of study, provide engineering students with linguistic competence sufficient to attend engineering courses and/or to serve an engineering internship in the target language.

A vital feature of the GEC model is that international engineering exposure is "built-in", providing a core level of international exposure for all engineering undergraduates. Although direct experience abroad (e.g., study abroad, international internship) is clearly the most desirable training, there will always be students who lack the required motivation to go abroad, regardless of how easy, streamlined and well-integrated the access to such opportunities. Such students, however, still participate in internationalized courses in our college, and must work on teams with international students via VSX. Thus, the Global Engineering College paradigm does not simply provide streamlined access to foreign culture and international engineering practice for a few students, but increases global awareness and experience for all students.

Supported by an NSF Department-Level Curricular Reform Planning Grant, we developed and piloted the four key elements of the GEC concept introduced above during the 2003 calendar year, based on our existing partnerships in Europe. Curricular modifications were focused on our award-winning Design4Practice program [18, 19, 20, 21], a unique practice-oriented engineering curriculum built around a four-year interdisciplinary sequence of design courses beginning in the freshman year, that incrementally expose students to design and teaming challenges of increasing complexity. The interdisciplinary nature of the Design4Practice program makes it very attractive to foreign students who often come from more rigid educational systems where interdisciplinary experiences are difficult to implement. Thus, Design4Practice courses provide an excellent foundation for the curricular internationalization and "virtual" international teaming elements within our GEC model.

Although all four of the elements outlined above were explored, the following sections focus on our experiences in implementing the key Virtual Student Exchange (VSX) element, as it is most relevant to the thematic focus of this volume. Section 3 describes the VSX concept, our analysis of requirements for a supportive VSX infrastructure, and our implementation of VSX tools. In Section 4, we describe our deployment of and experiences with VSX in a series of distributed team-based software design courses developed jointly with our international partners at German and Polish universities. In the final section, we briefly comment on future prospects for training students in distributed team-based design.

6.3   Virtual Student Exchange (VSX)

6.3.1   VSX: Concept and Motivations

As discussed in Section 1, modern engineers will increasingly be asked to work in globally distributed design and development teams as corporate internationalization and the 24-7 engineering paradigm continue to gain momentum. The only way that a student currently can gain experience in an international team is by going abroad for a term. Unfortunately, travel costs, semester timing, disruption of social relationships and curricular inflexibility often make such physical exchanges unattractive to students; an expensive study-abroad experience that delays graduation by two or more semesters is simply not practical in the eyes of many students. Of course, international exposure is only half of the solution; students also need training in distributed collaboration technologies and project management to work effectively in distributed design teams. The key to developing capable, well-rounded engineering graduates for the global marketplace, therefore, lies in developing a pedagogical model that efficiently provides for both international exposure and distributed teaming experience, while minimizing expense and curricular disruption in the student's degree program.

Over the last decade, sophisticated internet technologies have matured that allow us to provide a viable solution to this challenge. The goal of these technologies, collectively known as groupware systems, is to allow geographically distributed groups of collaborators to communicate, share data, and organize their collective team effort as effectively as possible, so as to minimize the negative dynamics of geographical separation. Although it appears unlikely that distributed collaboration will ever match the full efficacy of traditional (co-present) teaming [22], groupware technology has already proven to be quite useful: an increasing number of major corporations (e.g., Boeing, Ford, Sun Microsystems) have integrated distributed teaming using groupware tools into their business models to support design and development teams spread across widely-separated sites.

The concept of Virtual Student Exchange (VSX) was developed as a model for applying groupware technologies to provide distributed internationalized team-based design and development experiences in an educational setting. The overall idea is straightforward: apply the same groupware technologies currently being used extensively in modern corporate environments to allow students at partner universities to participate remotely in one another's design courses. Specifically, we have leveraged the VSX concept to enhance international exposure within our GEC curriculum in two ways:

1. Support virtual teaming. VSX supports the establishment of "virtual joint courses" centered around internationally distributed design teams. In this way, students who might never choose to go abroad are nonetheless exposed to international teaming, while at the same time learning to use a variety of groupware technologies to effectively operate in a distributed teaming scenario.
2. Provide a "virtual study abroad experience". VSX eliminates the logistical and financial obstacles associated with physically studying abroad by allowing a student to participate remotely in a desirable design course at a partner institution; optionally, he or she can then travel abroad at semester's end to physically join the semester in progress abroad.

In sum, the VSX model provides both training in distributed teaming tools/techniques and international experience by establishing distributed student teams spanning parallel design courses offered at participating partners institutions; students in such teams learn to use the same groupware technologies increasingly seen in modern industrial practice to coordinate the team's collaborative effort. Valuable experience is gained on all sides at minimal expense.

One problem we encountered (that plagues traditional study abroad programs as well) as we developed the VSX concept is dealing with the

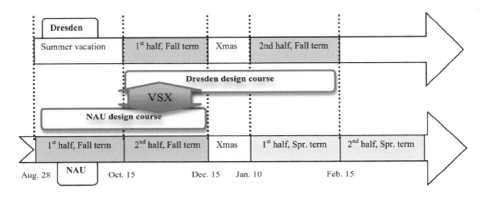

Fig. 6.1 Managing asynchronous terms under the VSX model.

asynchrony in semester timing that exists between different institutions: academic terms at one institutions often begin/end at different times than at a partnering institution. This complicates the establishment of "parallel" design courses at the two institutions, as the different start dates put the two courses out of sync. If the asynchrony is small (less than about two weeks), the problem is minimal, with each instructor making minor accommodations, and the two courses still running essentially in parallel as a joint course. For many American-European collaborations, unfortunately, the asynchrony will typically be quite large, as illustrated in Figure 6.1 for our collaboration with the Dresden School of Applied Sciences.

As in many European systems, the Dresden fall term starts relatively late — around the middle of October. In contrast, the term for American schools[1] typically starts on or about September one, resulting in a near half-semester asynchrony. Although an asynchrony of this magnitude rules out the possibility of truly "joint" design courses between the partnering institutions, there are still a number of quite fruitful alternatives:

- **Small project collaborations:** The most obvious alternative (illustrated in Figure 6.1), is to simply divide the design and development experiences within both courses into several smaller

---

[1] This applies to universities (like NAU) on the semester system; schools on a quarter system typically start later.

projects. Rather than collaborating on all projects (or one big one), distributed teams are formed up for one or more smaller projects that fall during the overlapping portion of the semester. Depending on the nature of the course co-opted for a VSX experience, this "partial collaboration" could actually be beneficial. In the junior level software engineering course we targeted in piloting the VSX concept, for instance, students were able to focus on other core SE concepts and skills taught in the course during the independent projects (i.e., in local teams), while focusing on the challenges associated with distributed teaming and project management during the joint projects.

- **An outsourcing model:** Design courses that are centered around a single large project can turn an obstacle into an advantage as well, by using an outsourcing model to integrate a VSX experience into the course. Outsourcing, of course, is a fairly common practice in modern engineering that raises many of the same communication and coordination challenges as full distributed teaming, as the "main" and "outsource" teams works to communicate requirements, coordinate joint development activities, and later integrate the work products of the outsourced segment of the project. A potential drawback under this model is that the "main" team enjoys a full large-project design experience, beginning with requirements acquisition, architecting the solution, and managing implementation and testing, while the "outsourced" team plays a somewhat more subservient role and completes a smaller project. This imbalance, however, can once again be turned into a feature by adopting an "apprenticeship" teaming model in which the "main" team consists of advanced students in their last semester, and the "outsource" team is drawn from a course that falls earlier in the degree program. In this way, (a) both teams gain the primary benefits of VSX collaboration, i.e., distributed team management experience in an international context, while (b) both teams do valuable work appropriate to their respective skill levels while gaining exposure to an important work model (i.e., outsourcing) commonly found in modern engineering practice.

- **VSX as a prelude to international exchange:** A final intriguing possibility is to leverage VSX to actually overcome the semester asynchrony obstacle as it pertains to traditional international exchange study. For example, consider an NAU student who wishes to integrate a study-abroad experience in Germany into her curriculum at NAU and, in particular, would like to participate in a novel design course offered in Dresden in the fall term. To do this, the student would normally have to skip the fall term at NAU in order to travel to Dresden for the start of the course there in mid-October — and then miss the next NAU semester as well in order to complete the course. VSX could offer a flexible alternative: the student would take her normal NAU courses in the fall, but also sign up for the course in Dresden, using VSX to participate as a solo "distributed" team member on the German design team. Following the end of the term at NAU, she could then travel to Dresden over the break, physically joining her team for the last months of the project. Again, this sort of "mixed mode" collaboration — partially at a distance, but with in-person periods of team interaction as well — is not uncommon in modern industry, where team members may occasionally meet physically for a "sprint" or other mission-critical segment of a project.

As this discussion illustrates, there are many options for integrating VSX-based international teaming experiences into existing engineering curricula, ranging from a fully joint course taught in parallel at both institutions to more flexible models in which two independent courses are connected for some level of distributed team development. The VSX model provides curricular designers with a powerful new tool for designing curricula that leverage the unique expertise and educational opportunities available within a consortium of partner institutions, while providing an efficient mechanism for training future engineers to operate effectively in distributed teaming scenarios.

In the following section, we turn our attention to the principles and analytic processes we used to determine the appropriate suite of groupware technologies needed to support effective distributed teaming in our educational scenario.

6.3.2 Designing Effective Infrastructure for VSX Deployment

Our exploration of the VSX concept can be divided into three distinct but overlapping efforts: analysis and planning, infrastructure implementation, and deployment/evaluation in joint design courses with our international partners.

Our analysis effort focused on identifying appropriate groupware functionality to support distributed design teams. A wide variety of groupware tools have been explored over the past 15 years [23]. More recently, a growing number of commercial groupware solutions have appeared (e.g., Skype, Onevision [25]; and Intranets [26]). Our analysis proceeded in two complementary directions: First, we identified the functional needs of student design teams [27, 28]. Drawing on the experience of our Design4Practice program faculty as well as our background in groupware systems development, we analyzed the activities and conventional mechanisms (e.g., project management software, intra-team data flow, email, etc.) used by traditional co-located student design teams to organize and coordinate team activities. This provided a set of minimal functionalities that any software solution would have to meet. Second, we examined a wide variety of existing groupware solutions, evaluating each on cost, ease of installation, and functional completeness. The outcome of this analysis yielded the following results, organized by functional category:

- **Communication tools:** Group members need to be able to communicate freely about the evolving design. After examining the needs of design teams, we concluded that email messaging should be sufficient to support intra-team communications. Although we considered real-time communication tools (e.g., text chat, video-conferencing) as well, there did not appear to be enough functional justification for real-time communication in our team design context. As will be seen later, this assumption turned out to be premature.
- **Access to designed artifacts:** In general, providing high-quality access to the artifact (e.g., a robotic toy, a solar array, etc.) being designed to all distributed team elements is extremely challenging, since the artifact being produced typically exists physically at only one "production" site. To limit this challenge, we focused our pilot effort on supporting VSX for computer science teams. Being

essentially text, software can be easily shared in a distributed context (although version control becomes an important issue).

- **Coordination and project management tools:** This area proved to be extremely challenging due to the diversity of coordination and project management tools and approaches that exist. Our analysis of local student design teams within our Design4Practice program revealed that most conventional student design teams maintained, in electronic form (e.g., project management software, electronic document) or on paper in their project notebooks:

    i. An evolving schedule for the project, with major tasks timelines somehow denoted.

    ii. Some mechanism for documenting evolving "to-do lists" for each team member was also common, ranging from a document on the team website to a whiteboard in the team's project room, to scribbled notes from team meetings.

    iii. Finally, team members frequently archived (at various levels of completeness), emails from teammates and notes from team meetings.

Under the assumption that coordinating team activities and keeping members on-task would be even more difficult in a distributed teaming context, we emphasized sophisticated calendaring, flexible task lists, and work status monitoring mechanisms in our survey of existing groupware applications. Note the emphasis on groupware applications, i.e., only applications that would support shared, distributed access by multiple group members was considered. After extensive evaluation, we determined that no existing commercial solution available at that time (2003) met our criteria. Although some products did provide shared calendaring functionality of some sort, no existing groupware solutions supported team task management: assignment of tasks to individuals, shared monitoring of task completion status, and capture of design notes/rationale. In addition, many commercial groupware offerings were based on a "subscriber model", in which a commercial "host" company maintains control and ownership of the group site, charging members a fee to use the groupware feature. This model is (a) not economically feasible for relatively low-budget educational contexts and (b) does not provide groups with the flexibility to customize their group space to the group's specific needs. For products based on the conventional "purchase and install locally" model, we found that (a) the installation process was generally quite complex, requiring a trained systems administrator to create and configure a shared group workspace and (b) the group environments provided were monolithic and inflexible, often overloaded with features irrelevant to team design contexts.

Based on this analysis, we elected to draw on our expertise in groupware design to create a custom groupware tool specifically designed to support small, distributed design teams. In addition to the functional requirements noted earlier, we set ease-of-use as a central design constraint, meaning that all components must be cheap, small, and easy to install. This constraint is dictated by both practical and financial realities: student design teams work together for relatively short periods (i.e., several months), so investment of substantial energy in a lengthy, complex setup is not justifiable. We also expect little funding for dedicated systems or support personnel, so student teams should ideally be able to do most of the configuration on their own.

6.3.2.1 The MOGWI System

Our software development effort resulted in an elegant groupware prototype called the Modular Groupware Infrastructure [35] (MOGWI) system that explores a novel ultra-lightweight, highly-modular groupware architecture. Key features of the system include:

- **"Thin client" architecture:** MOGWI does not require users to install any client software on their desktop machines. Rather, the MOGWI client consists of an applet-based core infrastructure that automatically downloads to the user's machine when the MOGWI website is accessed. This maximizes ease of use, while providing for easy universal access: group members may access the group site anytime, anywhere, from any machine with a web browser.
- **Flexible, Customizable, Extensible:** MOGWI is based on an innovative nested applet design, in which the initially down-loaded applet establishes a framework that hosts and provides core networking services to an extensible set of functional modules; these functional modules implement various groupware tools, e.g., project scheduler, virtual shared disk, team newsgroup, and so on.

Because the architecture is completely modular, individual teams may select and install only those modules that they feel they need to support their team activities on a specific project. New modules can be implemented and added to the running system at any time, becoming immediately available for teams to use.

- **Easy site maintenance:** Installing the MOGWI server software and managing groups in MOGWI is extremely simple as well. An install script allows a system administrator to install and configure the MOGWI server and database in about 30 minutes. After that, group management is trivial: to support a new design team, a MOGWI administrator creates a new MOGWI group and designates one team member to be the group administrator. This team member then adds all of his or her teammates, and configures the tools (modules) that the group should have access too. A design team can create and configure a collaborative work environment in less than 20 minutes.

After exploring a number of module concepts, we settled on four core MOGWI modules to implement for our pilot effort: an Awareness module, a NewsPost Module, a Task and Workflow module, and a Filebrowser module.

Figure 6.2 provides a snapshot of the MOGWI workspace with the four modules open; individual users can organize their view of the MOGWI workspace as they see fit. The Task and Workflow (TAW) module provides the core project management functions for the distributed team, by laying out project tasks on a Gantt chart timeline. Tasks may be created by any group member; responsibility for each task is assigned (by percentage) to some combination of group members; task completion status is maintained by the assignee. The TAW also supports task dependencies and thus can display tasks in a dependency graph, as an alternative to the Gantt view. In this way, all group members can easily determine what tasks remain to be done, who has been assigned to do them, how each task is progressing, and how their assigned tasks influence other tasks remaining to be done.

The NewsPost module implements a secure, flexible archive for group communications. Group members can initiate new topics (threads) of discussion, post new messages, and reply to existing postings. A sophisticated permissions mechanism allows users to specify whether threads and messages

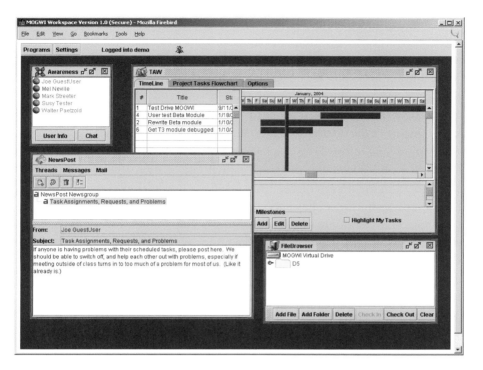

Fig. 6.2 The MOGWI workspace showing the four key modules.

can be read by others, replied to by others, or deleted by others, supporting a wide range of configurations. Another important feature of NewsPost is that it can be configured to serve as an automatic archive for emails sent to the group mail alias (which MOGWI also provides).

The FileBrowser module is essentially a shared virtual hard drive, a place where group members can store and make available arbitrary electronic files. Although the primary purpose of the Filebrowser module with respect to our pilot effort is to support sharing of software code being developed by our teams, it can also be used for picture files, design documents, and so on. To help manage coding projects, the FileBrowser supports a rudimentary version control system, allowing group members to check-out/check-in stored files.

The Awareness module represents our nascent attempt to support some "peripheral sense" of how hard (or whether) teammates are working, what they are working on, and some sense of "working together", which has been shown to be a critical factor in bonding a set of individuals together into a robust team [29, 30]. The Awareness module provides real-time feedback about which teammates are currently online, as well as a historical record of when and for how long they logged into the group workspace.

Combined with access to email, FTP servers, and other mundane internet technologies, the MOGWI system provided comprehensive support for the key elements of the distributed software production process that was the focus of our pilot VSX effort. The emphasis on social as well as practical considerations (e.g., the awareness module, explicit task assignment and completion status, etc.) in the design of our VSX infrastructure is important to point out. Distributed teams do not have access to the wide range of incidental information (e.g., peripherally noticing teammates in their cubicles working, a casual progress report in the hallway, etc.) that co-located teams take for granted, but which provides subconscious "glue" that holds a team together and generates commitment to the team. MOGWI makes a conscious effort to provide this key social information in some form for distributed teams.

6.4   Deployment of the Vsx Model: Lessons Learned

To evaluate the VSX concept and our MOGWI infrastructure, we integrated a variety of distributed team design experiences into two existing courses within our Design4Practice curriculum. "EGR286, Introduction to Engineering Design" is a sophomore level interdisciplinary design course team-taught by faculty from all engineering disciplines and focused on interdisciplinary design to meet a complex robotics challenge. "CS386, Software Engineering" is a traditional computer science course centered around formal software design, architectures and tools, and software project management; in our program, students explore these concepts hand-on, via a series of team-based software projects.

Drawing on existing contacts, we recruited collaborators in the computer science departments at two European partner institutions — the University of Applied Sciences in Dresden [32], Germany, and Wroclaw University of Technology [31], Poland — to help set up student software development teams spanning all three institutions. NAU took the role of the "hosting institution", meaning that NAU was responsible for designing the joint projects, establishing the goals and guidelines used to organize team interactions, and recording and evaluating outcomes.

Responsibility for developing specific projects to be tackled by distributed teams in each VSX trial was shared between all three international partners, with the selection shaped by the particular curricular priorities of each partner. Each partner would provide a set of ideas as a starting point, with subsequent discussion winnowing and reshaping proposals until a final version of the project emerged. As discussed in Section 4.2 below, we came to discover that, while the project itself was shared, the context in which the project appeared could vary between institutions. For instance, what was a networking class for one team (Germany) could be a software engineering class for another team (US). This flexible approach allowed for considerable autonomy in what specific material was presented and how it was presented at each institution.

Apart from the considerable freedom in determining topic and content of the course in which the VSX project was embedded, each partner also enjoyed general autonomy in how project contributions of local sub-teams should be evaluated; all teams had dedicated local faculty as supervisors, who took the responsibility of assessing performance for the students at their sites. Although this flexible evaluative model was necessary to accommodate differences between faculty, course foci, and, indeed, evaluative philosophies within different educational systems, it did require some refinement to ensure equal "motivation" to contribute to and excel at joint team tasks across all team members.

6.4.1 Initial Experiences and Insights in Deploying VSX

EGR286 was selected to host the first VSX-based distributed design pilot in Fall 2003. With faculty serving as pseudo-realistic clients for a regional emergency services agency, students were grouped into four "corporations" competing to develop a robot capable of exploring a debris-littered terrain after a disaster of some sort. Specific functional goals for the robots produced included navigating around obstacles, sensing the temperature of objects ("victims"), and delivering small aid packets. Student teams for each "corporation" were interdisciplinary, made up of sub-teams from computer science, mechanical engineering, electrical engineering, and civil engineering; the four corporations designed solutions in parallel, with a competition at the end.

While the other engineers designed the chassis, sensor systems, and testing terrains, student on the computer science (CS) sub-teams within each corporation were tasked with developing control software to allow the robot to navigate the terrain, mapping and avoiding obstacles (including other robots) and identifying "disaster victims". Robots could take bearings using radio beacons and process that data to derive positional and spatial information, and to construct an internal representation of the terrain and obstacles.

Adopting the outsourcing model described in Section 3.1, we established two remote teams in Dresden and Wroclaw that acted as "outsourcing consultants" to NAU teams. The local NAU CS teams were tasked with the core control interface for the robots; each team then "outsourced" the design of a sophisticated "smart" module that would allow robots to map obstacles and automatically return to base to the remote teams. Because each team in Dresden/Wroclaw served as an outsourcing contractor for multiple design teams at NAU, they were required to develop the requisite software with custom modular interfaces to integrate equally well with either of the implementations developed by the NAU-CS teams they were consulting for. Collaboratively specifying, testing, and integrating this outsourced module represented the main challenge for students in the VSX pilot.

Although the EGR286 course ran successfully, significant shortcomings in the design of the VSX element were apparent. The overall problem was that our implementation of the "outsourcing" model for collaboration did not adequately motivate the NAU and remote teams to collaborate as much as hoped. In particular, the failure to place specific evaluation emphasis (i.e., points in the course grading structure) specifically on communication, collaboration, and effective management of the outsourced teams allowed local teams to successfully complete the project without interacting extensively with their international partners. As a result, products produced by the remote teams were poorly integrated, often as an afterthought to the core project solution. Specific observations we made related to the quality of distributed team interactions include:

- **Weak NAU-international interactions:** Although there was considerable initial interaction between teams, it was not well organized, i.e., individual team members emailed back and forth a number of times, relaying vague requests for information. Interaction tapered off over time, particularly once the international teams felt they had enough information to proceed towards some solution on their own. In short, we found that students were not able to construct a truly effective collaboration when merely presented with a collaborative challenge and given communication channels; more structure is clearly required.

- **Lack of Synchronous Interaction:** Our initial requirements analysis placed a low functional value on synchronous communication channels, e.g., video-conferencing. It soon became clear, however, that we had greatly under-valued the social value of such interactions in generating trust, commitment, and team cohesiveness. In the latter half of the project, we hastily arranged for a video connection so that remote teammates could watch the testing and final competition of the robot prototypes. Simply seeing a remote teammate and chatting real-time appeared to greatly increase the excitement, commitment, and level of satisfaction of participants.

- **Poor use of software tools:** The MOGWI system was under-utilized. Several teams made use of the group mail alias, the Filebrowser (for sharing design documents), and the NewsPost module (to post design discussions). The Task and Workflow module received little usage, reflecting the poor coordination between local and remote team elements in distributing tasks and monitoring their completion. MOGWI usage was highest at first, and then tapered off as design and testing became more intense. Although this reflects the general tendency in student teams to ignore perceived "non-productive" (e.g., documentation, communication) tasks when time pressures increase, such lack of communication is particularly damaging when remote collaborators are depending on it to make progress themselves. An exacerbating factor here may have been frustrations caused by inefficiencies and bugs exposed in the MOGWI prototype (although these were generally fixed in short order).

In sum, our initial VSX effort helped to flush out a number of challenges associated with integrating distributed teaming experiences into our curriculum. We were particularly surprised to note that these challenges centered less on the technical shortcomings of the groupware infrastructure, and more around subtle social, motivational and team dynamics issues. Aside from a few minor technical flaws, teams were generally quite able to communicate and organize their collaboration with a remote partner, but simply chose to focus on work within their local groups instead.

6.4.2 VSX: Overall Outcomes and Observations

To investigate these issues, we engaged in a further series of extended VSX-based distributed teaming collaborations [33] with our international partners in subsequent semesters. Each of the trials was fundamentally similar to the one described in the previous section, but explored a number of variations on the theme, including:

- **Nature and level of targeted course:** Our initial pilot focused on the interdisciplinary EGR286 robotics design course, in which a central challenge for students (even in the standard co-located format) is organizing and managing communications between disciplinary sub-teams. To factor this issue out, and to investigate the potential effect of growing maturity/skills in the cohort, we tested the VSX concept in our junior-level CS386 (Software Engineering) course.
- **Nature of the collaborative relationship:** In our initial pilot, the remote teams were seen as "outsourcing suppliers", i.e., an entity "outside" of the core team per se. Speculating that this difference in status may have contributed to the weak local-remote communication, we explored a more centralized organizational structure in which local and remote members were truly on the same team with equal status and leadership mandate.
- **Different project designs:** One observation from our initial pilot was that, as deadlines approached and the stress levels within the project increased, communication with non-local team members flagged because it was not viewed as "absolutely necessary"; local sub-teams could make independent progress (even though this often led to integration problems later). To address this shortcoming, we attempted to design/constrain the project specifically to force close collaboration between the remote and local teams, e.g., by creating modules or sub-tasks that simply were not independent and would thus force closer collaborations.

- **Different evaluative emphases:** We also explored evaluative models (i.e., grading schemas) which placed significant value on team communications: team members were asked to document their communicative diligence with remote members, and submit this as part of the course grade. Although this was clearly an artificial constraint/incentive, it did allow us to generate enough communicative volume to properly assess the groupware infrastructure we had developed to support VSX.

As a detailed dissection of each of these VSX trials is beyond the scope of this chapter, we confine ourselves to a summary discussion of key overall observations and insights relevant to applying the VSX model in modern engineering education. The following paragraphs summarize our key observations and lessons learned.

6.4.2.1 Lesson: Course structure and organization is more vital than content

Our implementation of VSX courses explored a variety of course topics, project types, and teaming models. Participating successfully in international team collaboration is clearly a novel and demanding challenge for students and, indeed, often becomes the major challenge within a course, overshadowing the mastery of regular course content. With this in mind, our experience shows that the academic level or topical focus of the course is less important than making sure that the course is a suitable vehicle for teaching distributed teaming. In particular, we found that project-oriented courses are well-suited targets for VSX implementation, providing compact modular learning experiences, and a natural motivational context as teams work (and even compete) to produce a quality solution to the problem. Projects also provide a natural mechanism for introducing highly motivating realism into a course, with pseudo-realistic (or even actual) corporate sponsors to serve as "clients" for the project team. Where possible, projects should cover the complete design-implement-test cycle rather than, say, rote implementation of a functional specification. Not only is design an important skill to learn in its own right, the design phase in particular emphasized robust communication between team members, as the team works to determine and refine requirements, and come up with a basic solution architecture.

In sum, the actual topical content of the course — whether software engineering, robot design, game production, etc. — is not critical, so long as suitable design-and-build projects for collaboration can be articulated, and the parallel courses set up to support and vigorously exercise the distributed teaming skills that are the main focus of the VSX experience. Moreover, it is not even particularly important that the parallel courses at partnering institutions have the same topic focus. It is possible, for instance, for students in a course on computer game production at one institution to team with peers in a software engineering course at a partnering institution. As the teams work together to produce some software product (i.e., presumably a game, in this example), one half of the team is learning about game design and implementation at its home institution, while the other half is learning about and applying software engineering techniques and principles at its institution; both teams explore and apply their topical knowledge in the context of the software product being jointly developed. This provides substantial flexibility for institutions to serve their own curricular needs by maintaining a certain degree of local responsibility for the topical content taught in the VSX course, while at the same time exposing students to VSX-based distributed teaming.

6.4.2.2   Lesson: Design projects to require maximum interaction

In our first pilot VSX offering in the EGR286 course, we took a cautious approach by purposefully organizing distributed teaming as a loosely-coupled collaboration: the project was organized to permit collaboration with the remote sub-teams — but both groups could also largely succeed on their own. This approach was motivated by our desire to control risk of catastrophic failure (in a curricular sense) for students, by removing collaborative success from the "critical path" for the project. What if the groupware infrastructure we had put in place turned out to be inadequate? What if one sub-team or the other was less skilled or diligent than the other? Our initial instinct was that teams should not be "punished" by shortcomings beyond their immediate control.

This turned out to be the wrong approach, as noted earlier. Once the structure of the project became apparent and tasks were doled out, the co-located sub-teams quickly recognized that success (as defined by the course grading criteria) on "their part of the project" could substantially be achieved without extensive collaboration — at which point communication between the distributed team elements dropped off precipitously. In short, the project was structured so that communication and team interaction were not vitally important — and so students efficiently minimized this superfluous effort.

This problem was explicitly avoided in subsequent VSX trials: projects were chosen with a number of distinct sub-tasks, as before, except that now the sub-tasks were highly-integrated and mutually-dependent. In one course, for instance, the teams were asked to create software based on the cutting edge "Service-Oriented Architecture" (SOA) paradigm. The American half of each team was responsible for developing the server side of the product (i.e., the service), while the international sub-teams were responsible for the service requestor side (i.e., the client) of the architecture. The tighter coupling of the project parts paid off: students started communicating early and kept communicating regularly throughout the whole project.

The lesson here is that all sub-teams collaborating in a distributed team design project must be placed in a critical-path role, mutually dependent on one another in order for the project to succeed. This approach creates true motivation and strong incentive for communication between the distributed team elements.

6.4.2.3 Lesson: Provide robust, redundant frameworks for communication

We have already seen that projects must be designed specifically to require strong and continual communication within the distributed team, and have discussed how an appropriate choice of project is important to stimulate strong intra-team communications. The infrastructure provided for communication — both technological and social — is vital as well.

Based on the extensive needs analysis presented in Section 3.2, we were confident that the groupware infrastructure that we had put in place to support VSX (i.e., the MOGWI system augmented by email and other off-the-shelf tools) could cover all of the core communication and coordination needs of a distributed team. Although one could say that this was technically true, our practical experience showed that the dynamics of distributed communication are more subtle. In particular, our observations suggest that (a) an effective groupware infrastructure must — even in purely task-oriented scenarios like ours — carefully support not just functional communication requirements, but social needs as well; and (b) distributed communication is far more fragile and prone to (often permanent) breakdown than communication within co-located teams, requiring explicit compensatory measures. The following paragraphs explore these points in more detail.

6.4.2.4　Establishing and maintaining vital intra-team dynamics

When a new project begins in a co-located teaming context, there is generally some sort of "kickoff meeting" in which team members are introduced to each other and the project is formally initiated. There are certain rote functional motivations for such an introduction (e.g., identification of team members, exchange of basic skill set information, sketch of project plan, etc.) but it appears that such a meeting has a very high social value as well, serving to put faces to names, to demonstrate (through personal real-time presence) commitment to the project, and to develop an initial basis of trust within the team. Simply sending out an email to all members stating, "Welcome to the team! Here is the link to the MOGWI team site for our collaboration" simply is simply not nearly as compelling socially. In our later trials, we took a number of extra measures to establish a strong sense of group purpose, and to maintain this explicitly along the way:

- **Require and supervise a "live" kickoff session:** Our results were best with a video link between the distributed team elements, but telephonic conference calls or a multi-participant instant messaging environment could work nearly as well. The point is that (a) there is a formal event to kick off the collaboration and (b) that everyone is "present" in real-time to offer some sort of introduction and engage in casual discussion with peers. This explicit real-time "ice breaker" event seems to provide a basis of mutual knowledge and commitment that overcomes the natural reluctance to engage with complete strangers, and enabled subsequent asynchronous communication.

- **Provide video-conferencing where possible:** In most academic and corporate contexts, email has become by far the most common means of communication and coordination. We observed this as well in our analysis of co-located teams (see Section 3.2), and thus assumed that email would satisfy this core communicative need in a distributed scenario. Although email did play a central role in team communications, something was clearly missing. Emails to remote collaborators were typically quite lengthy, formal, and relatively infrequent; they were essentially modeled as formal letters. Moreover, emails from distant team members often were not answered promptly, if at all. There was also a pronounced tendency to blame remote team members if something went wrong, or to speculate about potential incompetence "at the other end". In contrast, emails between co-located team members were typically short, frequent, were generally answered immediately, and tended to be mutually supportive; these exchanges more closely resembled the sort of fluid, ongoing exchange common among friends.

The reasons for this are not particularly mysterious: not only did the co-located team members often know each other in advance of the project, but they all share a similar schedule, cultural background, and educational experience. In short, a broad range of social factors clearly plays a key role in generating a sense of "espirit de corps" and commitment to the team's shared goal. The important question for us was how to overcome (or at least ameliorate) the effects of geographic separation to create a similar dynamic in a distributed team.

In sum, our observation is that a real-time video connection appears to be quite effective[2] at creating a strong social connection between distributed teammates. When start-of-project kickoff meetings were held via a point-to-point videoconferencing systems, team members were much more likely to subsequently communicate by email, and generally seemed to promote a more positive image of remote team members. To keep the communicative momentum going throughout the semester, we built two further mandatory video-conferences into the curricular plan near the middle and end of the project. While these three "all-hands" videoconference meetings were the only mandatory group meetings, individual team members were also free to utilize the videoconferencing workstations as needed on their own.

After investigating a number of videoconferencing tools available at the time (2005), we ultimately settled on VRVS34 as the video conferencing

---

[2]Presumably, a face-to-face meeting to start the project (as is common in corporate contexts) would be even more effective, but this is not realistic in most academic scenarios.

system. VRVS allows for multiparty videoconferencing as opposed to a one-to-one connection provided by many of the instant messaging systems. Apart from an inexpensive camera and microphone VRVS doesn't require any additional resources. The software is free.

Even though VRVS is a simple system, the effect on the students was amazing: communication with remote elements of the teams became a more natural, fluid part of the project, and students appeared to be much more motivated to provide the best quality of service possible, responding to requests and task assignments in the same prompt fashion observed in the co-located teams.

6.4.2.5 Plan for communication breakdown

In co-present communication, participants have a large arrange of resources available to monitor the efficacy of communication and keep it on track. For example, visual access to a conversational partner can give cues as to whether an utterance was not heard, was misunderstood, or is being purposefully ignored; this access allows participants to interpret unfolding events properly to avoid misunderstandings and quickly detect and repair communicative breakdowns. All of these peripheral cues, however, are missing in asynchronous email conversation, and this led to substantial problems in our early trials. A particular problem was what could be called a communication deadlock: when email or a groupware tool did not work reliably or its operation was not fully understood, a "mutual waiting" situation would often result, in which each participant is waiting for the other to act. We have all experienced this situation at one point or another in email-based conversation: a message is sent that clearly begs for a response...but then that message is either not delivered (e.g., due to a transient technical failure), leading to a deadlock as the first participant waits for a response from a second participant who never received the request. A similar situation occurs when the request is received, but the response is somehow not delivered. In co-located scenarios, there are multiple resources to repair this breakdown: participants either know each other well enough to risk "nagging" with a follow-up email, or they meet each other in a hallway and mention the open request. Distributed teams have few "repair" mechanisms of this sort; in many cases, these deadlocks were only resolved through explicit intervention by an instructor to "restart" the dialogue.

Although such situations can never be totally avoided in the generally asynchronous context of distributed collaboration, the problem can be mitigated considerably by setting up a formal "communication protocol" made known at the start to all distributed team members. In our case, the protocol consisted of two simple ground rules that all teams were expected to adhere to:

- It is expected that all messages will receive a response within a short timeframe; the exact time depends on the differences in time zone between the team members, but is generally understood to mean "almost immediately after receiving the message at the start of a working day".
- In case a full answer cannot be given immediately (e.g., the question is complicated and requires extensive research to answer), at least a confirmation message was required within the period of time indicated by the first rule.

The effect of this simple protocol was, in essence, to formalize expectations and thereby to define a basis for assuming communication failure: if no response was received within the given timeframe, the sender would have reasonable cause to assume something had gone wrong and to repeat the query. More generally, this experience illustrates our point that a communication infrastructure designed to support distributed teaming must anticipate the potential for breakdown, recognize that the distributed work context severely limits the resources available for detecting and repairing such breakdowns, and must provide appropriate compensatory mechanisms.

6.5 Discussion: VSX Best Practices

As evidenced by many of the other projects and initiatives described in this collection, distributed 24-7 collaboration models for research and product development are becoming more and more common in modern global industries, as global communications connectivity increases, prices for bandwidth fall, and competitive pressures push companies to tighter development timelines. At the same time, the evidence is quite clear that working effectively in distributed teams is substantially more challenging than in co-located teams; new skills and work techniques are required. A central goal of the VSX element integrated into our Global Engineering College model for internationalizing engineering education is to expose students to these challenges and help them gain these critical distributed teaming skills as a regular part of their undergraduate education.

Although our initial explorations of the VSX concept have been promising, our experiences have also shown that — just as working effectively in distributed design teams is more difficult and trouble-prone than one might imagine — organizing effective learning experiences to teach distributed teaming skills is particularly challenging as well. In particular, our observations indicate that the success of international teaming experiences in higher education is very sensitive to how an internationalized class is prepared and managed; even small weaknesses can lead to a deterioration of productive interaction. As a learning experience, the pilot VSX efforts provided many insights — both expected and surprising — on the implementation of international teaming in design courses. Some of the "best practices" we have drawn from our experience include:

- Choose a project task that can be divided into different but closely coupled parts to inspire maximum collaboration and interaction within the distributed team; all elements of the distributed team need to be placed in a critical role, so that each team element can only succeed by communicating with other (distant) elements.
- All parties have to understand and "own" the project. Thus, develop the project to be done in close collaboration with all partners.
- Effective distributed teaming does not necessarily mean lock-step parallel coursework within participating institutions. Participating institutions may retain substantial autonomy over the local class's contents, so long as the collaborative segment is topically relevant and meaningful to students within all participating local contexts.
- Similarly, partnering institutions retain local control over student assessment, so long as (a) substantial assessment emphasis is placed on the volume and quality of collaboration between distributed team elements, and (b) there is some fundamental agreement as to metrics for measuring and assessing quality of outcomes.
- Explicit organizational measures can be taken to encourage strong communication:
    - Start initial contacts very early and apply administrative pressure to ensure that it happens. Videoconferencing

seems to have a particular social value in establishing team identity and commitment in initial team meetings.

○ Monitor asynchronous communications carefully for volume and quality. We have found that "communication archives" established as a regular piece of the team's work documentation are an unobtrusive way to monitor communication traffic. Set up rules to prevent communications from getting stale.

• Groupware tools are essential to the success of VSX-like training experiences. However, make sure that the tools are well understood and work reliably. Concentrate on the functionality that is really needed.

6.6 Future Prospects for Training Students in 24-7 Team Design

Our experiences with VSX have been overwhelmingly positive, and we plan to build on these experiences to further refine and extend our VSX offering in the future. We also plan to explore further technical refinements as a way of enhancing the VSX learning experience. Although the MOGWI system provided a satisfactory starting point as an integrated backbone for distributed team collaboration, it is clear that it could be much improved by leveraging recent exciting new developments in web technologies. For instance, AJAX, in-browser drag and drop, and other Web 2.0 techniques moving the web towards browser-based deployment of fully functional applications promise to offer better performance (and lower programming complexity) than our current nested-applets approach. In addition, the landscape of commercial or freely available communications tools has changed substantially since we began the project; systems like Ventrilo, Skype, and Elluminate provide higher reliability and an impressive range of functionalities that could be selectively integrated to improve our evolving VSX infrastructure. The cost of implementing and maintaining MOGWI, in both effort and time, was substantial; avoiding such costs by integrating off-the-shelf communications solutions where possible has obvious advantages.

A more subtle lesson learned from our experiences with VSX is that, as an instructor, organizing and offering a successful VSX experience takes substantially greater time and effort than offering the same team project experience in a conventional co-located context. First, extensive advance communication and coordination with collaborating faculty at distant institutions to carefully plan the project and how the team elements will interact in its execution requires a large investment in up-front effort. Next, the technological infrastructure must be put into place and tested before the collaborative segment begins; and finally, oversight of the project as it moves along required more effort as well, as the faculty member monitors intra-team communications, scrambles to iron out break-downs and technological glitches, and continues to coordinate with colleagues abroad. Altogether, we estimate that administrative effort required of participating faculty ranges between 150–250% (depending on size and nature of project) of what might be expected in a conventional setting. Thus, there is a very real danger of "burn-out" among the motivated faculty that is the life blood of successful pedagogical innovations like VSX. It is absolutely essential that administrative elements in participating locales recognize this increased faculty load, and account for it appropriately in local faculty workload and evaluation models.

Finally, we are excited by the prospects for rejuvenating and expanding the VSX and GEC concepts inherent in the increasing visibility and perceived importance of internationalization in modern engineering education. The number of schools interested in networking and participating in VSX-like projects is growing rapidly, driven also by the increasingly global market for engineering education in general. For example, our university has joined a consortium of institutions developing innovative dual-degree programs (in our case, a $1 + 2 + 1$ program with China) under which international scholars complete part of their studies at a home institution, and part here at our university to earn dual degrees from both institutions. Integrating VSX into such models could not only help overcome curricular and semester timing issues between institutions, but could also provide critical training in distributed international teaming at the same time.

6.7   Biographical Information

- *Eck Doerry* is an associate professor in computer science at Northern Arizona University with research interests in groupware systems and online collaborative communities. He is a long-time advocate of internationalization of engineering education, and in this context, has explored a number of tools for supporting collaboration in internationally-distributed design teams.

* *Wolf-Dieter Otte* is an assistant professor in computer science at Northern Arizona University with research interests in distributed systems' technologies; his most recent project focuses on exploration of advanced portal technologies to support the Arizona Water University initiative. His experience and interest in international education led him to lead implementation of the VSX joint courses.

References

[1] P. Altbach, "Perspectives on International Higher Education," (Resource Review column), Change, 34(3), 29, 2002.

[2] K. Gray, G. Murdock, and C. Stebbins, "Assessing Study Abroad's Effect on an International Mission," Change, 34(3), 44, 2002.

[3] M. F. Green, "Joining the World: The Challenge of Internationalizing Undergraduate Education," Change, 34(3), 12, 2002.

[4] S. Marginson, "The Phenomenal Rise of International Degrees Down Under," Change, 34(3), 34, 2002.

[5] D. Maxwell and N. Garrett, "Meeting National Needs: The Challenge to Language Learning in Higher Education," Change, 34(3), 22, 2002.

[6] M. Miller, "American Higher Education Goes Global," Change, 34(3), 4, 2002.

[7] J. Sjogren and J. Fay, "Cost Issues in Online Learning: Using 'Co-opetition' to Advantage," Change, 34(3), 52, 2002.

[8] T. Bikson and S. A. Law, "Global Preparedness and Human Resources: College and Corporate Perspective", Rand Corp, 1994.

[9] D. Lambert, "The Winds of Change in Modern language Instruction," International Educator, 31–35, 57, 1999.

[10] Oregon State University, International Academic Program http://oregonstate.edu/international/studyabroad/degree.

[11] University of Rhode Island, Engineering Program, http://www.uri.edu/iep.

[12] V. Ercolano, "Globalizing Engineering Education", ASEE PRISM pp. 21–25, 1995.

[13] F. L. Hubbard, "Innovation Is the Key." ASEE Prism, 13(5), 5, 2004.

[14] D. McGraw, "My Job Lies over the Ocean." ASEE Prism, 13(4), 24–29, 2003.

[15] D. McGraw, "Putting It in Perspective." ASEE Prism, 13(5), 24–29, 2004.

[16] CEFNS International Certificate Program, http://denali.cse.nau.edu/CETIC/Intnl Certificate.html.

[17] Noir sur Blanc, "Survey of Engineering Studies Worldwide," Paris, France, 2000.

[18] K. Collier, J. Hatfield, S. Howell, and D. Larson, "A Multi-Disciplinary Model for Teaching the Engineering Product Realization Process." 1996 Frontiers in Education Conference, Salt Lake City, UT, 1996.

[19] J. Hatfield, K. Collier, S. Howell, D. Larson, and G. Thomas, "Corporate Structure in the Classroom: A Model for Teaching Engineering Design." Proc. of the 1995 Frontiers in Education Conference, Atlanta, GA, 1995.

[20] S. Howell, T. Harrington, D. Larson, and G. Thomas, "A Virtual Corporation: An Interdisciplinary and Collaborative Undergraduate Design Experience." 1996 Design for Manufacturability Conference, Irvine, CA, 1996.

[21] D. Larson, "A New Role for Engineering Educators: Managing for Team Success." Proc. of MRS Spring 2000 Conference, San Francisco, CA, 1999.

[22] E. Doerry, "An Empirical Comparison of Communicative Efficacy in Technologically-mediated Environments". Dissertation. University of Oregon, 1995.

[23] R. Baeker, "Readings in Groupware and Computer-Supported Cooperative Work: Assisting Human-Human Collaboration." Morgan Kaufman Publishers, 1992.

[24] Netmeeting, see http://www.microsoft.com/windows/netmeeting/.

[25] OneVision, see http://www.ingenux.com/onevision/.

[26] Intranets, see http://www.intranets.com.

[27] Y. Rogers, "Coordinating Computer-Mediated Work", Computer Supported Cooperative Work: The Journal of Collaborative Computing, 1(4), 295–315, 1993.

[28] R. Vick, "Perspectives on and Problems with Computer-Mediated Teamwork: Current Issues and Assumptions." ACM SIGDOC Journal of Computer Documentation, 22(2), 3–22, 1998.

[29] P. Dourish, and V. Bellotti, "Awareness and Coordination in Shared Workspaces." Proceedings of ACM CSCW'92 Conference in Computer-Supported Cooperative Work, Toronto, Canada, 1992.

[30] C. Gutwin, S. Greenberg, and M. Roseman, "Staying Aware in Groupware Workspaces." Proceedings of the ACM CSCW'96 Conference on Computer Supported Cooperative Work, Boston, MA, 1996.

[31] Wroclaw University of Technology, see http://www.pwr.wroc.pl/en_main.xml.

[32] Hochschule für Technik und Wirtschaft Dresden, see http://www.htw-dresden.de/

[33] W.-D. Otte, W. Paetzold, and J. Nicodem, "Virtual Student Exchange in the Global Engineering College", Proceedings of the 2005 American Society for Engineering Education Pacific Southwest Regional Conference, April 2005, Los Angeles.

[34] VRVS (Virtual Room Videoconferencing System), see http://glast-ground.slac.stanford.edu/workbook/pages/getting_connected/vrvs.htm.

[35] MOGWI, see http://ccl.cefns.nau.edu/Groupware/mogwi.

7

Data and Knowledge-Transfer Model for the Development of Software Requirements Analysis CASE Tools designed for Cross-Time-Zone Projects

Zenon Chaczko*, Bruce Moulton†, Jenny Quang‡, and Karan Jain§

School of Computing and Communication, Faculty of Engineering and IT, University of Technology, Sydney, Australia
*zenon@eng.uts.edu.au, †brucem@eng.uts.edu.au,
‡jenny.quang@uts.edu.au, §karan.jain@.uts.edu.au

Abstract

This article describes work undertaken to evaluate an approach for developing collaborative requirements-analysis CASE tools that are specifically designed to address the needs of cross-time-zone development teams, that is, teams spread across different geographical locations around the world. Few of the software requirements analysis computer assisted software environment (CASE) tools readily available are designed specifically for cross-time-zone development activities. We propose a specifically tailored data and knowledge-transfer model, and investigate its suitability for the development of a cross-time-zone oriented CASE tool. The approach was used to develop a working prototype. The approach and prototype will be further evaluated in a collaborative undertaking involving the Wroclaw University of Technology, the University of Technology, Sydney and the University of Arizona (UA).

Keywords: Knowledge transfer, Requirements analysis, CASE, Cross-time-zone

157

7.1   Introduction

It is well known that doubling the size of a team does not halve the development time. To reduce development time, organizations have increasingly been adopting a practice which makes use of additional teams located in various spots around the world. This "24 hour" mode of working is commonly found in open source development projects, and increasingly used by large companies. While one team sleeps, another can continue the development during its daylight hours. The Open Source community (For example the Linux kernel and Apache web server projects) has worked in this fashion for years. Many large organizations including IBM, Sun Microsystems, Cisco Systems, Nokia and Google also use geographically and temporally spaced development teams [1]. Twenty four hour continuous development is ideal for tasks that have hard-deadlines or require work completed as soon as possible. If a functional or security bug is discovered in a mission critical application, there is a need to find a solution within the shortest period of time. For example, the approach might enable a "three day solution" to be completed in a 24-hour period. The two day difference might be extremely valuable in terms of down-time costs.

However, the process whereby teams work in different locations has significant effects on the way that the work is structured and organized. For example, the project leader is not available to the "night" team. Each team must work independently of the other, and each must hand over the work to the other at the end of the shift.

The emerging trend in cross-time-zone development is sufficiently prevalent that several curriculum developers are exploring ways of enabling students to gain experience in this mode of working (e.g., [2, 3, 4]).

It is proposed that the development process might benefit from efforts to acknowledge and convey both explicit knowledge and tacit knowledge. Tacit knowledge can be understood as uncodified knowledge that leaves when employees leave the project). This problem is ordinarily handled by attempting to "convert" tacit knowledge into codified knowledge, by way of documentation. However prior research suggests certain aspects of tacit knowledge can only be transferred through face-to-face contact [5]. Nonaka and Nishiguchi suggest that most knowledge is created not by individuals, but by interaction and dialogue among several people [6]. Distributed teams have limited opportunities for face to face contact; hence such knowledge transfer issues are particularly relevant for cross-time-zone development projects.

Prior work suggests that certain environments have specific characteristics that render formal methods of knowledge transfer inadequate. For example, one study suggests that some engineering sites experience difficulties during attempts to codify/document certain aspects of their more experienced employee's knowledge for simulation or formal training purposes [7]. It has been proposed that inability to transfer knowledge can be a hazard where there are safety critical operations, and that this must be taken into account during the design of workflow processes [8, 9]. Brown and Duguid suggest that knowledge transfer is less facilitated by converting tacit to declarative knowledge than by aligning the goals and practices of employers and employees [10].

A vast range of computer assisted software environment (CASE) tools can be used to assist knowledge management during software development. Many of the tools focus on providing support for a range of complex abstract concepts and representations, for example, UML, SDL, Z-spec etc. A characteristic ordinarily exhibited by such tools is that the complex levels of functionality can cause the tool to be a hindrance to split teams where each hands over work to the other at the end of each shift. At the time of writing, very few of these tools appear to be specifically designed to support cross-time-zone software development, and this was the motivation for evaluating a data/knowledge management approach for the development of two types of tools for managing software requirements in multi-time-zone development: Requirement Analysis and Management Front End tool (RAMFE) and Software Requirements Analysis CASE tool (SRAC).

7.2 RAMFE Case Study

Requirement Analysis and Management Front End (RAMFE) is a CASE tool which enables a project team to operate around the clock; it is next leap to current project management front ends. Not only does it allow you to manage project attributes but it also acts like a central point of reference for the whole project team. Its benefits range from project details to iteration management and delivery. RAMFE is based around AGILE methodology allowing customers greater flexibility with their requirements and allowing development teams to focus their delivery on specific features. Some fairly simple yet very effective concepts have been adopted in this engineering tool. The front end supports a complete project life cycle, it allows the project manager to initiate a new project and add sub projects. Add project team, stakeholders and assign roles to individuals.

The concept aims to aid project management and change management activities. There were 2 distinctive phases of development the RAMFE concept; In Phase 1, the generation of behavioural tree was completed and in Phase 2 I the Behavioural tree was integrated with project management and change management components. The front end is in its conceptual phase with a prototype available.

7.2.1   Change Management and Project Management

The application holds critical information highlighting the sponsors, the project manager, the due dates, the list of project and initialised sub-projects under it. The tool provides users with a set template to enter the requirements, this leaves no room for ambiguity and when searching for specific require-ments, the job is a lot easier. Communication is also a main emphasis; the front end is to aid the entire project team member to be in well informed about the progress and their involvement with the active tasks in the project. It aids business analysts and project managers to detect flaws in requirements earlier and reduce the risk of deliverable.

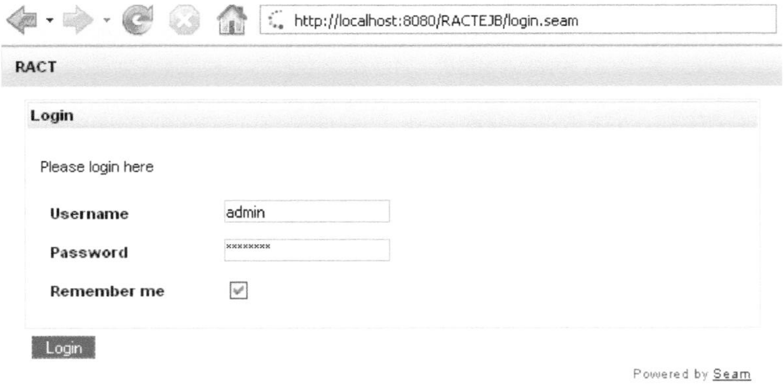

Fig. 7.1 RAMFE login entry.

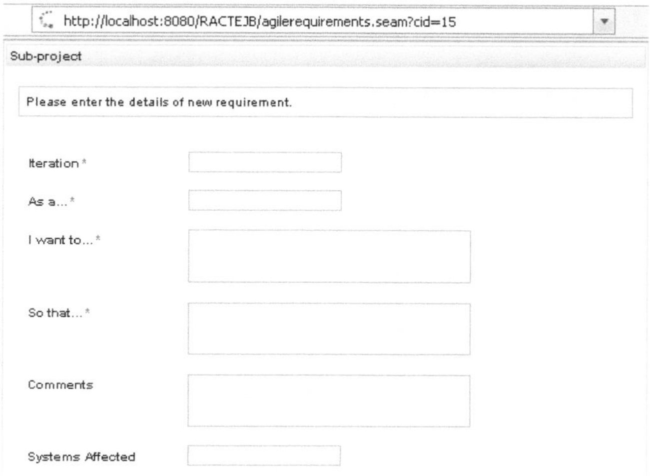

Fig. 7.2 RAMFE subproject entry.

Requirement Analysis and Management, RAMFE has a requirement analysis tool embedded in it. With user intervention, RAMFE can transform requirements into a two-dimensional behavioural tree. The tree is hierarchical and the path from the root of the tree to a particular feature. The emphasis of this feature is to help software engineers and business analysts understand the requirements clearly and understand the impacts on other parts of the project. Functional requirements exhibit behaviour usually with an Actor or an Event triggering another action. The Requirements Analyser aims to help Business Analysts detect errors in their requirements by analysing the semantics of each requirement sentence to ensure their correctness. It provides the ability to model the requirements in the form of a graphical behavioural tree that can then be used by Business Analysts or Engineers to detect requirements errors or discrepancies early in the Software Development Lifecycle.

7.2.2 Export Requirements and Document Repository

One of the most beneficial functions for this application is its ability to export the requirements to excel and submit them to SVN (Figure 7.3). This can potentially allow users worldwide working on a project is able to share the most recent version of requirements and collaborate as a successful team. The

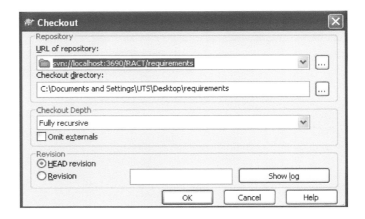

Fig. 7.3 RAMFE export requirements and document repository entry.

document repository can be operated using any SVN open source tool for collaboration by project team members. The current prototype has distinguishable components to it:

- Requirement capturing
- Requirement analyser
- Exporting the requirements to excel
- Submitting the requirements to SVN
- Basic project management

7.3 The SRAC Case Study

A primary goal was to investigate the suitability of a data/knowledge-transfer model for the development of a SRAC tool. The work was to focus not on just the analytical parts of the documentation, but rather the entire shape of it. It was envisaged that the SRAC tool should be suitable for diverse skill sets, and suitable for the support of requirements analysis involving a minimum of overhead while facilitating effective document standardization and sharing of information (requirements artifacts). The process was to draw from IEEE Recommended Practice for Software Requirements Specifications (IEEE 830-1998) [11]. The tool was intended to provide a framework for standardization of system/software requirements documentation at both local and global levels,

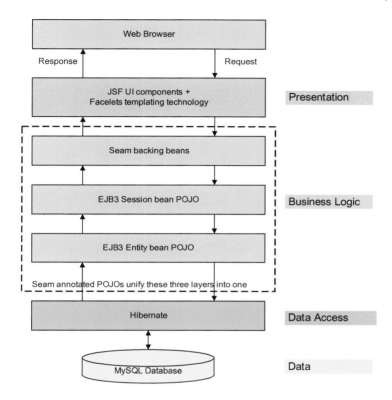

Fig. 7.4 A layer model of RAMFE.

and at the same time remain a shared data repository that enables exchange of information across different locations, time zones, system development environments and documentation formats.

The approach included considerations relating to iterative and incremental development processes that would be suitable for a cross-time-zone collaborative development environment. It was proposed that such teams would benefit from a structure for documenting small iterations in development (8 hour shifts) and methods for allowing for periodic resynchronization. It was envisaged that it would be valuable to permit a way for inter site issues arising from shift to shift to be identified, documented and perhaps isolated and planned to be rectified by the same or successive shifts. An agile software development methodology known as Scrum was chosen, in part because it focuses on managing complex development processes iteratively. Issues

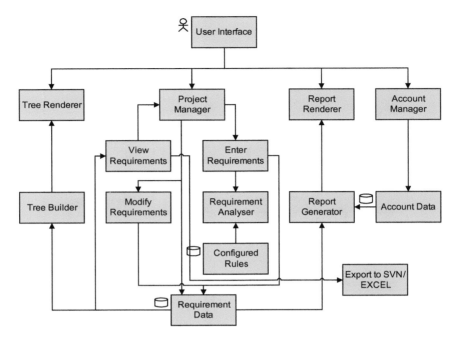

Fig. 7.5 A conceptual architecture of RAMFE.

relating to handover-synchronisation can be handled using the Scrum process skeleton. From a project management perspective, it was proposed that the Scrum methodology may assist in synchronizing intensive development tasks.

It was proposed that it would be useful from an organisational knowledge point of view if data (for example files) that was produced during a shift, and if subsequent alterations to this data, could be captured and correlated against work done in previous shifts. The approach viewed the modifications to data as being similarly noteworthy as the data itself. This approach is different from existing tools, which over a long period of time record only the persistent data of the project. It was noted that decision making in distributed teams can also be fragmented, and individual teams may make decisions that affect the entire project and must be recorded and distributed to all teams.

To handle the above considerations we introduced the concept of Eventflows. The Eventflows concept is an adaptation of the Lifestreams concept coined by Freeman and Gelernter [12]. Where Lifestreams record the digital events of a single person, Eventflows record the digital events of a project

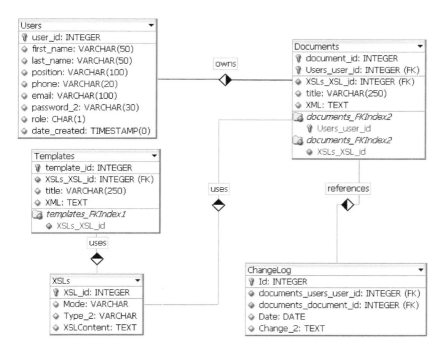

Fig. 7.6 Entity Relationships Considerations relating to Data/Knowledge-transfer model.

and the project artifacts. Eventflows capture events and periods within the project's global system, and capture and distribute project knowledge (Figure 7.1). Events were classified as any significant occurrence on the project that can be captured or recorded by a development environment, for example the login or logout of a system, the commit of changes to a version control system or the modification of a project artifact. A period is the linking of two key events where on their own has little or no value. Eventflows can be captured through automated systems or through manual creation from users. Eventflows can also be linked against individual or groups of tasks defined in project management tools such as Microsoft Project, thereby showing the actual work that was required and accomplished to complete the task.

7.4 Evaluation of the Approach

Considerations relating to the Eventflow methodology led to the following attributes/constraints: (1) an Eventflow must consist of both human readable

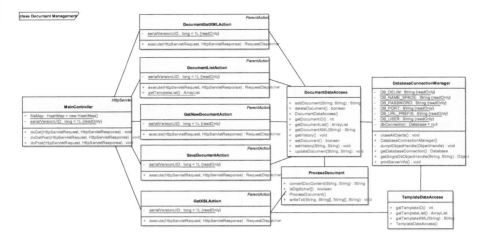

Fig. 7.7 Diagram regarding document management, arising from considerations relating to the data/representation model.

and application readable data; (2) it must include date and time of creation; (3) it must include a human readable description of event; (4) it must include a machine readable description of event (implemented as serialized object); (5) it must include a project identifier; (6) it must include an artifact identifier.

The evaluation suggests that the database design plays a crucial role in the design of a SRAC, because it decides what data will be stored in the system, what type of user queries will be easily provided and how the rest of the system will interaction with the database. Key data persistent components identified during the evaluation of the approach include: (1) account information and details pertaining to a user are stored in a table, and each user is identified by a unique key; (2) templates are stored table, each has a unique id, entire templates are kept in text/XML format, each has a reference to the XML stylesheets in an 'XSLs' table; (3) documents created from templates are stored in a 'Documents' table, and documents are kept in text/XML format and updated each time the document is revised. Like the templates, each document has a reference to the XML style-sheets in the 'XSLs' table; (4) to capture the historical changes made to an existing document, when a change occurs, it is logged in a "Change Log" table. Each entry is identified by a unique identification number and also holds the identification number of the document it corresponds to; (5) templates and documents are kept as XML in the data tables. XML is human readable, however it is not aesthetically formatted for viewing. To provide styles and formatting for XMLs, the 'XSLs' table contains XSL style sheets.

The evaluation considered diagrams providing views of document management and associated entities, including those given in Figures 7.1 and 7.2. The relationships between significant components of the system architecture can be best described using collaboration diagrams, and the basic dynamics can be demonstrated using high level sequence diagrams. For example, the Participants component is shown using collaboration diagrams, and its dynamic is visualised using in high level sequence diagrams. The approach was evaluated for its suitability for implementing views of various layers. The View layer contains all the components associated with presenting the user interface that allows the end-user to view and interact with the system. Of the two distinct interfaces that make up the interaction screens to the tool, first enables the user to manage the application users, and the second is for the creation, and viewing of the templates and documents. The Controller layer brings the model and view together and integrates the application. The Model layer contains the business logic and components that access the data in the database and manipulates the data. The Data View depicts the key persistent elements of the system.

For the purposes of this project, we estimated that the average number of users supported by the tool on the project would be 100, while the rate of document creation is 3 per day. (The expected volumes of traffic would vary depending on the type and size of the development project.) Even though the functions available within the tool are not strictly time critical, the performance of the system is still expected to be responsive to a user's actions. It was estimated that the system would easily respond to a user's action in less than 0.5 seconds.

The evaluation also considered the extent to which the approach was suitable for the development of a system that is scalable, secure, reliable and portable. The prototype was a web based application, so security was imperative to ensure that any confidential information and intellectual property is more accessed by any unauthorised parties. The prototype was secured via authentication and authorisation mechanisms. Following the MVC architecture will provide another level of security, where users cannot directly access specific sections of the CASE Tool. Reliability was also a key consideration because the prototype was intended to be available 24 hours a day. Portability was also a consideration so that developers in different locations could access the tool from different environments. The IEEE Recommended Practice for Software Design Descriptions (IEEE 1016-1998) [11] served as the basis for which the design description was written, and the design is customizable according to the particular attributes of the project. The design description describes the various classes to be built, how the database will be set up, what the system graphical user interfaces will look like, and what the interactions within the system are. Components relating to the View Layer of the system are made up customizable JSP/CSS files. The Template Management section of the system handles all the functionality surrounding the use of templates.

The approach was found to be suitable for producing documentation in accordance with IEEE standards for software design descriptions. The resulting tools appear to be suitably scalable, and a range of features can be added including uploading of ready documents, incorporating an integrated help, notation toolkit, and interoperability with other products/systems. Overall, the evaluation suggested that the data and knowledge-management approach had successfully led to the development of a simple and effective prototype.

7.5   Conclusion

This article considers a new approach for developing software-requirements analysis CASE tool specifically intended to suit the particular needs of cross-time-zone development projects. The approach was evaluated in light of a data/knowledge-transfer model. The model is intended to enable developers to focus on issues relating to the codification and transfer of critical events and knowledge within and between development teams. The preliminary evaluation suggests that the model is suitable for informing relevant development-related considerations. This finding is consistent with the prior research. However further work is required to explore the limits of the model and determine the extent to which the model is appropriate for the development of tools that are scalable for large numbers of users.

References

[1]   A. Gupta, "Expanding the 24-Hour Workplace", The Wall Street Journal, September 15, 2007.

[2] Z. Chaczko, J. D. Davis, and C. Scott, "New Perspectives on Teaching and Learning Software Systems Development in Large Groups — Telecollaboration", IADIS International Conference WWW/Internet 2004 Madrid, Spain 6–9 October 2004.

[3] Z. Chaczko, R. Klempous, J. Nikodem, and J. Rozenblit, "24/7 Software Development in Virtual Student Exchange Groups: Redefining the Work and Study Week", ITHET 7th Annual Conference Proceedings, pp. 698–705, 2006.

[4] A. Gupta and S. Seshasai, Toward the 24-Hour Knowledge Factory, MIT Sloan School of Management. USA. Available at http://papers.ssrn.com/sol3/papers.cfm?abstract_id=486127, 2004.

[5] J. Roberts, "From know-how to show-how? Questioning the role of information and communication technologies in knowledge transfer", Technology Analysis & Strategic Management, 12(4), 429–444, 2000.

[6] I. Nonaka and T. Nishiguchi, Knowledge Emergence: Social, Technical, and Evolutionary Dimensions of Knowledge Creation (New York: Oxford University Press), 2001.

[7] B. Moulton and Y. Forrest, "Accidents will Happen: Safety-critical Knowledge and Automated Control Systems," New Technology, Work and Employment, 20(2), 102–114, 2005.

[8] B. Moulton, Enabling Safer Design via an Improved Understanding of Knowledge-related Hazards; A Role for Cross Disciplinarity Australasian Journal of Engineering Education [in press, accepted 22.12.2008], 2009.

[9] B. Moulton, Conventions to achieve safer design and reduce catastrophic and routine harm to the environment 2009 International Conference on Environmental and Computer Science (ICECS 2009) Singapore 22–24 January 2009.

[10] J. S. Brown and P. Duguid, "Structure and Spontaneity: Knowledge and Organization", in I. Nonaka and D.J. Teece (eds), Managing Industrial Knowledge: Creation, Transfer and Utilization (London: Sage), 44–67, 2001.

[11] Institute of Electrical and Electronics Engineers, (1998) IEEE Recommended Practice for Software Design Descriptions (IEEE 1016-1998), IEEE, New York.

[12] E. Freeman and D. Gelernter, "Lifestreams: A Storage Model for Personal Data", SIGMOD Record, 25(1), 80–86, 1996.

Index

RIVER PUBLISHERS SERIES IN INFORMATION SCIENCE AND TECHNOLOGY

Other books in this series:

Volume 5
Biomedical and Environmental Sensing
J.I. Agbinya, E. Biermann, Y. Hamam, F. Rocaries and S.K. Lal (Eds.)
November 2009
ISBN: 978-87-9239-28-8

Volume 6
Pattern Recognition and Machine Vision
Patrick Shen-Pei Wang (Ed.)
March 2010
ISBN: 978-87-92329-36-3